I0609159

Aristotle on Life and Death

Aristotle on Life and Death

R.A.H. King

Duckworth

This impression 2004
First published in 2001 by
Gerald Duckworth & Co. Ltd.
90-93 Cowcross Street, London EC1M 6BF
Tel: 020 7490 7300
Fax: 020 7490 0080
inquiries@duckworth-publishers.co.uk
www.ducknet.co.uk

© 2001 by R.A.H. King

All rights reserved. No part of this publication
may be reproduced, stored in a retrieval system, or
transmitted, in any form or by any means, electronic,
mechanical, photocopying, recording or otherwise,
without the prior permission of the publisher.

A catalogue record for this book is available
from the British Library

ISBN 0 7156 2982 4

Typeset by Ray Davies

Contents

Contents

Abbreviations and conventions

Metaph.	*Metaphysica* (*Metaphysics*)
EN	*Ethica Nicomachea* (*Nicomachean Ethics*)
MM	*Magna Moralia*
EE	*Ethica Eudemia* (*Eudemian Ethics*)
Pol.	*Politica* (*Politics*)
Rhet.	*Ars Rhetorica* (*Rhetoric*)
Poet.	*Poetica* (*Poetics*)
Frg.	*Fragmenta* (*Fragments*)

Other abbreviations

LSJ	H.G. Liddell, R. Scott and H.S. Jones, *A Greek English Lexicon*, 9th edition. Oxford 1990.
DK	H. Diels and W. Kranz, *Die Fragmente der Vorsokratiker*, 3 vols, 6th edition. Berlin 1951. (Presocratic thinkers are quoted and cited using DK's numbering.)
FHSG	*Theophrastus of Eresus, Sources for his Life, Writings, Thought and Influence*, edited and translated by W.W. Fortenbaugh, P.M. Huby, R.W. Sharples, D.M. Gutas. Vol. 5: R.W. Sharples, *Sources on Biology*, Text 1992, Commentary 1995. Vol. 3.1: *Sources on Physics*, Commentary 1998. Leiden.

Conventions

Translations are my own, unless otherwise noted. Additions in translations are marked by square brackets. Greek words, even when quoted, are not placed in quotation marks.

Acknowledgements

This book began as a doctoral dissertation: my primary debt is to my supervisors, Professor Myles Burnyeat and Dr Robert Wardy. Professor Sir Geoffrey Lloyd was also unfailingly generous with time and advice. My examiners for the PhD, Professors Malcolm Schofield and Bob Sharples, read and commented on a complete draft of the book. The Classics Faculty at Cambridge contributed to my fees and living expenses for the first two years of work; the British Academy awarded me a doctoral studentship for the second two. After finishing my thesis, I worked on Professor Thomas Buchheim's Deutsche Forschungsgemeinschaft project in Mainz on the concept of nature as the foundation of objectivity in Aristotle. Both he and Dr Johannes Hübner, a colleague on the project, have provided inspiration during the book's preparation, not least by reading a draft of it. Professor Mary Louise Gill also performed this latter service, and so saved me from many mistakes. An anonymous reader for Duckworth read the book and made a number of helpful suggestions. The Indexes are largely the work of Katharina Luchner. Deborah Blake at Duckworth has been extremely supportive and patient.

It gives me great pleasure to express my gratitude for all this help. Remaining mistakes and omissions are, of course, my responsibility.

Munich, 2001 R.A.H.K.

1

Life-cycles

1.1 Mortals and others*

[Apollo] tore down the wall of the Achaians
very easily, like a boy his sand-castle by the sea,
who, when he has made the plaything in childish whim,
tumbles it down again with hands and feet in play.
Iliad 15.361-4

A child playing with a sand-castle is a good image for Apollo's easy
demolition of the Achaean fortifications at Troy: construction and destruction imposed on entities unable to resist. A sand-castle's structure is
simple, pretty arbitrary; it is an artefact, and above all it is passive in the
face of its surroundings; its existence obeys no rules of its own. But an
image can illuminate also by its weaknesses, its failure to correspond to
its correlate. For natural things, living things, can be described in a
preliminary but nonetheless fundamental way by using the contraries of
the predicates that apply to the sand-castle: they are complex, active and
tightly organised.

Nonetheless, similarities remain – the castle has a shape or form, as do
living things, and its shape is in some stuff. These similarities might seem
to be undermined by the differences: the way that an animal has a shape,
the way the shape is in something, is radically different, not due to the
activity of an external agent, but to the animal's own activity, its growth.
This point suggests a further strength of living things: their activity
protects them from destruction. Limpets and suchlike thrive under conditions which would prove disastrous for a sand-castle over time. On the one
hand, of course there are structures, such as shells, which protect living
things from their environment; on the other, the whole animal, including
its shell, is produced by that animal's activity. Castles do not repair
themselves, do not replace their matter in the course of nature. Rather,
they are dependent on their makers.

Contrast the sand-castle with its maker: the latter is a living thing, the
former is not. The child was born, is growing, one day will be mature, will

* This title is taken from a collection of Bertrand Russell's essays.

1

then age and will finally die. His existence follows a regular life-cycle. He was produced by parents of the same kind as himself; but after birth he grew and will decline quite of his own accord, even if he remains healthy; nothing makes him do it. This is not to say that these changes are uncaused; merely that the (decisive) causes lie in the child's nature, or rather the nature of which he will be an example when he is mature. Humans, like other living things, run through a life-cycle. Throughout these changes, it remains constant that the child has to nourish himself (eat and breathe, for example). A wide variety of other activities may be performed on the basis of this process.

However, one may wonder quite how independent living beings are. They too can be destroyed by inimical surroundings, their stuff also comes from their surroundings, and they also fail in the end, if they are not Homeric gods. Their activity runs its course and then ceases: there is a limit to their ability to preserve their activity. And a limit to their activity is a limit to their independence. But it is a question of how and why there is a limit to their activity – is it something inherent in that activity, or in the stuff they are made of, or is it imposed from outside? – and perhaps there are other possibilities. But it is clear that to understand how the activity is limited, we must understand the activity – that is, how it takes place in living bodies.

This book attempts to give Aristotle's account of these facts about living things. I begin with a brief consideration of other treatments of the subject (§1.2), pointing to the central phenomenon of nutrition, and then locate Aristotle's treatment and preoccupations relative to a modern considera-tion of life-span; relevant above all to Aristotle's treatment is the idea that by running through life-cycles, living things ensure their permanence, as contrasted with the modern preoccupation with a mechanism timing the stages of life (§1.3). The subject is a natural or physical one; and indeed it touches the heart of Aristotle's conception of *physis* (nature), as he con-ceives this quite generally: nature is a principle of change, and the central case of this is the growth of living things (§2.1). Such a principle is the form of the living thing that imposes itself on matter, representing what the thing is when it is mature, that is when it reaches its end (§2.2); as such the form, as determining what the thing is when fully grown, exists, and so serves as the clue for investigating living things (§2.3).

After this general view of living things, I proceed to the investigation that takes up the bulk of the book: that into soul and living things. Aristotle's views on living, ageing and dying are contained in his work on the soul, *de Anima*, and its completion, the series of small investigations commonly and perhaps unfortunately known as the *Parva Naturalia* (*Small Natural Treatises*) – more especially, *de Longitudine et Brevitate Vitae* (*On the Length and Shortness of Life*) and *de Juventute, Senectute, Vita et Morte* (*On Youth, Old Age, Life and Death*) (§3.1).[1] The first step, following Aristotle, is to give a preliminary definition of the soul, which marks off living things from everything else, as the primary actuality of a natural body with organs; this definition is then given concrete form in the functions or parts of the soul, but it is closely associated from the outset with the nutritive soul, that is, the soul responsible for nutrition, since all living things, in all their parts, must achieve this actuality or activity (§3.2). It takes place when the heat of the body functions as a tool working on food, turning it into that body and so preserving it. The actuality stays the same throughout life, but more or less food can be processed, thus accounting for growth and decline (§3.3). In accounting for longevity, the first question to be asked is whether what makes something easily perishable, and so short-lived, is its body or its soul. The answer is the body, for it is characterised by contraries and so changes and perishes (§3.4). But the soul is the soul of the body and so must be somewhere in it; a wide variety of locations is possible depending on the degree of organisation of the living being concerned, but all have in common, according to Aristotle, that they are in the middle; the extremities can be lost without loss of life, but vital activity requires an organ or part to perform the central function. Vital activity comprises nutrition in all living beings, and also perception in animals, a capacity constituted by the balanced heat preserved by nutrition (§3.5).

To understand this function we have to see that living bodies comprise 'mixtures', in which contrary capacities for change, hot, cold, dry, wet are in balance, and the organs or tools made of these mixtures (§4.1). The food needed to form and preserve living bodies is drawn from their surroundings, and there must be some balance between the surroundings and the living thing, such that the latter is not overwhelmed by the capacities for change in the former (§4.2). Longevity is then to be explained by the relatively greater heat, and the moisture this feeds off, in the relevant living thing; heat provides the explanation on the basis of its role in nutrition (§4.3). The central function is performed by heat, which must be cooled to prevent it burning itself out. Because heat is necessary for this function, living things must remain hot, and so cooled throughout their life (§4.4). The cooling is performed in a variety of ways in the various living things, at the simplest level by the intake of food and the action of surroundings, in insects through air inside them providing a cooling breeze (§4.5); in the most developed living things by organs – lungs and gills – which are full of hot blood cooled by an interchange with the surrounding air or water. The heat serving the form or vital activity of these things is part of their explanation, alongside, if subordinate to, their form (§4.6). Blood flows to the heart where it is concocted by the fire there (*pneumatôsis*), thus raising the chest, expanding the lungs and drawing cool air into the proximity of the hot blood in the lungs, which then cools,

3

sinking the chest. And so on continuously, as long as the cooling organs are able to perform their function; death occurs when they fail (§4.7).

This enquiry has been general, taking in all the classes of living things in Aristotle's view. Thus it allows him to generalise about the life-cycle and its stages – in other words, to define them. Abstracting from the particular ways in which cooling is performed, and the manner of nutrition, he can place the start of a new life at the first participation in nutritive soul, taking place in hot matter; such participation continues as long as there is life – increasing in youth, decreasing in old age – and ceases when cooling is no longer performed, causing death (§5).

So much for a lightning tour of our enquiry. A little more orientation is in order for the more contentious readings of Aristotle that it contains. The following is meant more to whet the appetite than satisfy it; and not all the fine things to come are among the appetisers. Some general features deserve mention to start with; even if they are not particularly contentious, they are unusual. First of all, it is to be stressed that we are dealing with an enquiry with a beginning and end, starting with the delimitation of the subject matter, and ending with definitions, even if those parts of *de Anima* and *Parva Naturalia* that are most commonly discussed have been left out, above all those on perception and thought. We combine those parts of the enquiry most obviously philosophical, because universal, with those apparently most physical, because particular, from the conviction that for Aristotle universality is to be achieved only via a general view of the terrain. The subject matter is living things, divided into kinds, but also viewed as one kind, namely as things possessing soul, but also as things with life-cycles. The general definition of the soul in *de Anima* applies to all living things; and the definitions of the stages in the life-cycle with which this study ends are explanatory in that they rely on an examination of the different classes of living things conducted in earlier chapters. On this basis the definitions possess a generality secured by an investigation of the kinds concerned.

I have been sparing in my use of other writings of Aristotle; important exceptions are two pieces from his general account of coming to be and passing away (*de Generatione et Corruptione*) on growth (see §3.3) and the constitution of living body (§4.1); as the name of the work suggests, both passages describe the role of change in constituting things. One other argument from that work (II 11) is also of crucial importance, namely that living things that are mortal are subject to life-cycles, since in this way they can be permanent features of the world. In turn this is based on the idea that it is good for things to exist, and existence in cycles is as close as finite beings can come to God's unvarying life. This argument is only touched on as a preliminary piece of framework to our real business, the analysis of living things. Although these are substances in Aristotle's view, his analysis of substance, the enquiry of the *Metaphysics* into what there is, is not our subject. So that mention of the *Metaphysics* is almost entirely confined to the notes, with the exception of the description of nature taken from the lexicon of terms, which has no exclusively metaphysical significance (§§2.1, 2.2).[2] Aristotle's physics has all too often been neglected in favour of his metaphysics, as though the latter could be understood in isolation.

The problems we face are staples of Aristotelian scholarship: understanding matter and form, and their relation to one another. Broadly my approach is not controversial in that both form and matter must serve as explanatory factors in Aristotelian physics. Unusual is my emphasis on the function of nutrition, and the insistence that living things nourish themselves throughout their lives, that is, literally continuously. As food-burners, they must burn food or be extinguished. Put in a different way, soul is an actuality, and not merely a capacity for any one of a number of living functions. The advantage of taking form in this way is that it can make clear how form exists in matter – by imposing itself on matter, and may only exist in matter – for without matter for it to impose itself on, it ceases to exist.

Since this actuality is the soul, it offers a guiding line right from the start of the enquiry, throughout the variety of living things, up to the definitions of the stages of life; this provides a relatively uncontentious view of Aristotle's teleology, here to be seen as the thesis that living things fulfil an end, involving both an ontological commitment as well a methodological role. However, in a broader perspective, reflections on method, and on the relation between what Aristotle is doing here and his theories in the *Posterior Analytics* would form the matter for another book.

While that traditional bugbear of Aristotelian scholars, mixture, has not proved any less difficult for me than it has for anyone else, at least one aspect has been rescued that has been too little emphasised: the continued presence of the capacities of the ingredients in the mixture. As capacities for change, they form the basis in one sense for the regulated changes in an activity. In the mixture, these capacities remain, although they are in something new. In this sense at least, Aristotle can be said to compose his activities from comprehensible ingredients, the problem that strikes us at least most forcibly when confronted with his conception of matter as continuous: it possesses no structure at a microscopic level.

Looking for the seat of the soul is perhaps no longer as fashionable a pursuit as it once was, but understanding the soul, as I claim Aristotle does, as an actuality of body, enables him to pursue this topic systematically. It is widely recognised that the enquiry into the soul in *de Anima* is intimately connected to the enquiry into actions common to body and soul in *Parva Naturalia*: and one way they are connected is in the fact that the latter locates the soul in a suitable part of the body. The soul has its seat in those parts or organs most closely associated with the actuality of nutrition and perception in the relevant living things.

The definitions of the stages in life, with which we end, highlight the ingenious way in which the living thing grows and decays, depending on the quantity of matter it processes. These definitions represent a high point in Aristotle's natural philosophy – definitions of life and death are rare in the literature. Our disappointment should not be too great when we see that life here is taken right from the start as being life that starts; on this basis, its end is not too difficult to describe. For as an activity in matter, the soul's cessation can be explained by considering the body necessary for the soul's activity; but the soul must still be mentioned in describing what is happening. What this means of course is that life with neither end nor beginning is not the subject of our treatises.

1.2 Interpreting Aristotle on life and death

Consideration of the nutritive soul, even in *de Anima*, requires mention of the bodies concerned, but the so-called *Parva Naturalia* are explicitly concerned with those functions of living things common to both body and soul. The final treatises in the series of investigations with which we are concerned, *de Longitudine et Brevitate Vitae* and *de Juventute, Senectute, Vita et Morte*, are often considered purely 'physiological' rather than 'psychological'. Today, we would not consider the explanation of longevity or of life-cycles as parts of psychology. But the idea of the nutritive soul makes all the difference for Aristotle. All living activities are related to this kind of soul, and so may count as psychology.[3]

When we turn to the literature on Aristotle, we notice the huge amount that has been said about perception and the small amount that has been said about nutrition.[4] This is, of course, hardly surprising, for the particular reason that the post-Cartesian view of the relation between body and mind is prominent nowadays; and for the general reason that philosophers have a tendency to understand life primarily in terms of cognition rather than the metabolism necessary for its support.[5] If they notice the phenomenon of feeding, it is mainly with regret; and with a warning about its dangers for the philosophical life.[6] Aristotle is different; the theoretical question of what living things are, is for him fundamentally about their self-preservation; if we do not understand that, we will not understand his theory of living things as a whole.

Despite the general neglect of nutrition, two scholars have recently offered interpretations of Aristotle's views on the nutritive basis of life – Gad Freudenthal and Mary Louise Gill.[7] Here, I wish to make some general points about their approaches as a way of introducing some problems. More detailed consideration will be given in the appropriate place. The problems in question are traditional staples of Aristotelian

6

scholarship – the question of the relation between soul and body, and the coherence of a science which is concerned with form and matter – and our chief resource is the explanatory potential of the nutritive soul. Freudenthal's interpretation has a distinctly traditional ring, for he thinks that Aristotle uses 'connate pneuma', the breath contained in the body, as the sole cause of the presence of life and death. Pneuma is, Freudenthal thinks, an otherwise mysterious substance which serves as the substrate for the heat of living things, their 'vital heat'.[8] Freudenthal concentrates on vital heat because the four elements (earth, water, air, fire) alone are not enough to explain the mortality and persistence of living things,[9] and he is looking for the explanation of vital activities in a 'material substance'.[10] But he is unable to connect this material cause to the activities it helps to explain, and is forced to divide Aristotle's theory of living things into 'two theories of life': 'physiology' and 'psychology'.[11] Because he cannot see both 'theories of life' as belonging to the one science of nature, he thinks that Aristotle's theory of living things falls apart, and is only half-heartedly held together by the incomplete and very obscure theory of pneuma.

An important feature of Freudenthal's work lies in the way he emphasises the heat of living things – vital heat. Less helpful is the idea that Aristotle is a vitalist – that is, that he makes a single stuff alone responsible for the presence of life. For what we want, and Freudenthal does not provide, is a comprehensible view of what Aristotle was doing. Mysterious substances do not provide us with rationally attractive views of Aristotle. Freudenthal's approach seems to imply that Aristotle was unable to account for the coherence of body and soul without introducing some special stuff to achieve it.

Once again, nutrition offers Aristotle the key. Not only is it a vital function that is obviously not viable without matter, it is also necessarily closely connected to form. The form remains by way of nutrition, whereas the matter serving this end is changed. The importance of the idea that living things are hot lies in the way they can be conceived as living off their nutrition like a fire. If mention of fire, along with matter flowing and a form remaining, reminds the reader of Heraclitus, then that is all to the good, for Aristotle is indebted to Heraclitus for these thoughts on the way change ('flux') can support a determinate form or *logos*. As we shall see, the idea is that the burning of the natural heat of a living thing – that is, the activity of its nutritive soul – converts food stuff into the stuff of the living thing. This burning preserves the natural heat of the living thing and provides it with the capacity to perform other vital functions.

Mary Louise Gill's interpretation of Aristotle on life and death arises largely from her investigation into his views on substance, particularly in the central books of the *Metaphysics*.[12] Her understanding of the soul is that it is an 'active capacity' that enables the living being to live a particular sort of life:[13] to nourish itself, reproduce and perceive. Furthermore, she thinks it is the soul, as a capacity to produce change ('active capacity'), which preserves matter, i.e. elemental matter, in a living body from decay. Soul also directs growth in living beings.[14] In contrast with these active capacities of the soul, she sees matter as responsible for decay: the living being has to exert itself continually against decay, because it is made of lower-level matter. This matter causes complex living things to decay, which she sees as a return to simpler matter.[15] On her view, Aristotle ascribes the primary responsibility for ageing and death in living things to their matter. She makes the elements or 'generic matter', as she calls it, ultimately responsible for decay in living things.[16]

Now, it is plausible to think that matter causes our destruction. Things like sand-castles tend to decay; and such things are material. And, in the same vein, matter is responsible for our life: we are causally dependent on our constituent matter. According to Gill, it appears that the proximate matter, i.e. the living body, of the living thing contributes to the living activity, and the matter out of which the body is composed also contributes to life; but it is this matter that in her view causes life to fail. So there seems to be room for an account of this ambiguous status of matter, enabling life to happen, but also ensuring its failure.

This thought leads on directly to another major problem with any theory of living things that posits an immaterial factor such as the soul: how can something immaterial have an effect on something material such as a body? On Gill's reading of Aristotle the soul is an active capacity, and one of the things it is meant to do is 'control' fire – that is, prevent fire from fulfilling its natural inclination to rise. It is very difficult to see how soul could do this without itself having a tendency to sink, as earth does. But if soul were earth, it would be a body. However, my thesis is that Aristotle understands soul as actuality – that is, basically as nutrition. Nutrition is the activity that turns food into bones, flesh and so on. So if this activity is the soul, we can say that the soul produces the living body and as such is not to be identified with the matter that it converts into the living body or with the living body itself. But it is clear that this will involve restricting the untrammelled exercise of capacities in the nourishing matter. And this process will involve not merely one informed stuff in the body, *pace* Freudenthal, but all the organs of the body, insofar as they contribute to its preservation, *and* insofar as they are preserved by nutrition. Nutrition is a complex activity involving the interaction of capacities for change in the living body.

One might ask: is death inscribed in soul or matter? Since these are the two internal aspects of living things, if death is to be understood as natural to them, then it would seem that death must be the result of one of the two. (Compare the demise of Homer's sand-castle.) Yet these are not the only possibilities: death might be caused or explained by both form and matter. This is clearly the case with life: explaining living things requires talking about both their souls and their bodies. So we would do well to consider life as something connecting body and soul: it is common to both. How can we conceive of life as involving both form and matter? One way would be to say that soul is responsible for life in that soul inscribes life into matter, that is, gives it the form of a living thing or its parts. This is nutrition, the assimilation of food stuff to the stuff of the living thing. Thus we can understand life from the way that soul, the living form, inscribes itself into matter. And, to turn back to death, just the same fact may be the explanation here too. For if soul exists by informing matter ('inscribing itself into matter') then the way this informing of matter happens might plausibly be thought to determine how long it can continue. The kind of body a nutritive soul produces for itself determines how long the living thing lives, and so explains its ultimate, natural demise.

Aristotle sometimes talks, not of the death of living things, but of their decay or perishing (*phthora*),[17] and this would suggest that he saw the decay of living things as part and parcel of the general tendency of things to decay. In other words, the sand-castle and its maker decay in the same way. And that would seem to make decay susceptible of a very simple explanation; namely that the matter (animate and inanimate) out of which things are made is subject to change which destroys the form present in the matter, so bringing about the end of the composite of matter and form. There would be no special explanation of the decay of living things, merely one general account for the whole sublunary realm. But this is not the way Aristotle proceeds; rather his account in our treatises involves an elaborate account of the actuality and organs of living things. So while we must admit that there is a connection between decay in the general sense and death, we cannot simply say that they are the same. All things have the capacity to change, which involves them in decay; and living things draw their matter from the world around them, and so are related to inanimate things. If at all possible, the change from inanimate to animate should be accomplished without the help of magic.[18] The account of nutrition must explain this change.

The lessons to draw are that life is to be understood primarily as an activity of the living body, not merely inherent in one part of the body. The nutritive soul can be seen as the burning of the natural heat of living things that assimilates food to them. This burning has to be seen in relation to matter – the body it produces, the nutriment it consumes. Thus nutrition is intimately connected with its demise. Insofar as life preserves or nourishes itself, death is not external to living activity, not merely caused by its clay feet, as it were. The idea of the soul controlling the body has been mentioned; and we now turn to this control and its relation to life-cycles.

1.3 Aristotle and the explanation of life-cycles

The growth that nutrition causes allows living things to run through a series of stages in their lives, what one could call their *life-cycle*. It is these stages in life – coming to be, youth, prime, old age, death – that Aristotle defines at the end of his enquiry into life and death.[19] As the idea of a life-cycle might suggest, Aristotle treats life and death together. The concept of life has, of course, a broader application,[20] but we will only be treating the life of mortal things, insofar as this is opposed to death. He thinks of death as the limit to life.[21] This is the reason they are treated together. In modern treatments of death, life is also treated for the simple reason that death is seen as the permanent cessation of life.[22] So treatments of death have the task of saying what it is that ceases – that is, the definition of death depends on understanding life.

Perhaps perversely, I wish to approach the idea of life-cycles through ageing and death, rather than birth and youth. What we are primarily concerned with here is death as a part of the natural life-cycle, not violent death. This perspective is one that is not common in modern discussions of death, which tend to be concerned with the value of death.[23] Nonetheless, the idea of a natural life-span has not been entirely forgotten.[24] For it is connected closely to the idea that life is lived from birth to death, and not the other way round.[25] Life-cycles provided Greek thinkers with a way of articulating the link between life and death.[26] Two closely connected ideas are involved. First, the life of an individual moves from birth to death through a succession of stages. Secondly, life and death follow one another in a cycle. We shall be almost exclusively concerned with the first idea.

One can distinguish between a criterion for death and the definition of death. The former is an aspect of living things allowing one to tell when death has occurred; the latter is a question of what death is, that is, of the concept of death. In traditional terms, death could be diagnosed through the permanent and irreversible cessation of lung and heart activity.[27] Nowadays, the criterion of death has become elusive on account of technical innovations allowing the artificial replacement of heart and lung function. So one may either speak of the cessation of the *unaided* function of lung and heart; alternatively, one may speak of 'brain death' – the failure of function in the brain. In fact the preferred practice is to use a combination of these two criteria.[28]

The question of these newfangled criteria need not concern us: they lie far beyond Aristotle's horizons. He orientates himself using heart and lung function; and because he sees the heart as the centre of sensation and consciousness, his view includes an analogue of brain failure. But, as one would expect given the distinction between criteria and a definition, his

10

definitions of life and death are not concerned with these functions. Considering that these definitions are meant quite generally, and that Aristotle does not think that all living things breathe, or have hearts, that is perhaps not surprising. One advantage of distinguishing between criteria and definitions of death is that even though Aristotle did not have the problems with criteria that we do, his treatment of the concept of death may still be valid.

The vital function which Aristotle uses in his definitions of the stages of life is nutrition:[29] roughly, new life is there when the thing can nourish itself, and as long as the process of nutrition is increasing, it grows; it reaches a peak when mature; and then it declines. When nutrition ceases, death occurs. A brief defence of the use of nutrition is in order here. More commonly nowadays, vital functions such as adaptation to the environment and self-replication are seen as the best candidates for defining activities of living things. But it would seem reasonable to think that to perform these activities, a living thing must nourish itself, and so provide itself with the energy and parts to perform these activities.[30]

Nutrition has been criticised as a suitable defining activity for life and death on the grounds that there appear to be living things which do not engage in nutrition – for example, certain moths have no mouths or digestive tracts.[31] Since such a moth is alive, and is not able to nourish itself, nutrition would seem to fail as the defining characteristic of finite living things. Such animals live only to reproduce and die, and are dependent on stored energy. But such singular counter-examples to the view that the presence of life is to be defined through nutrition should be regarded with caution. For such a moth is not an independent form of life, but rather one stage in the life-cycle of a kind of animal. That is to say, the energy from which such a moth lives must be understood in terms of consumption of food by the caterpillar from which the moth has developed. Thus although the moth itself does not perform all the parts of the nutritional function, it is not a different animal from the caterpillar which actually did perform the eating and digesting necessary for the energy to be stored, and indeed for the moth pupa to be formed. That the moth at some given time performs no digestion should no more worry us than does the lack of digestion on the part of a *very* hungry monk, although the monk could, should he so wish, feed, and the moth could not. The point is that both are dependent on past nutrition for their continuing life, and as long as they are alive, metabolism is taking place. The energy stored by the moth is being used by its metabolism; if there were no actual metabolism, the moth would be dead. However, there is an important lesson for us in the example of the moth, namely that nutrition can be divided into discrete parts. As the example shows, the different parts – in this case ingestion and use of energy – need not be simultaneous. But the important thing is that nutrition as a whole is not the same as its parts. Obviously, as long as

11

the moth is alive its living tissue is able to preserve itself and provide itself with energy; and this is the bottom line of nutrition.

Caterpillars and moths provide a striking illustration of an important characteristic of living things: life-cycles. Life-cycles are central to Aristotle's theory of living things, as befits someone interested in living things generally, and, more particularly, in their regularities. The control of growth and decay presents serious problems for someone, such as Aristotle, with no idea of the massive complexity of living things, let alone any idea of genes. One might describe his ignorance as lack of knowledge about the mechanisms causing the stages of life. But it is worth asking whether the stages in life require special mechanisms to trigger them.

The issue can be roughly stated as follows, if one takes ageing as an example of a stage in life.[32] A prominent worker in the field has suggested that there may be a so-called biological clock which sets off the process of ageing.[33] Biological clocks are physiological processes which not only take time but also measure it.[34] They form necessary parts of the explanation of many aspects of regular behaviour in living things. So one suggestion is that the failure of function in living things, while it may happen on many levels, is finally determined by some master clock that sets off the onset of ageing.

Another suggestion, not incompatible with the first, is that things age because, as their cells replicate and so preserve themselves, they become increasingly erratic: errors accumulate and tissues and organs become increasingly liable to various kinds of malfunction, finally ceasing to be able to function at all.[35] Errors may be due to a range of causes. The most basic source lies in the mechanism of cell replication but a variety of biological waste products has also been proposed as being responsible for contributing to the decrease in viability of the organism.

Obviously, in considering Aristotle's theory, there can be no question of such an *internal* time-keeping device.[36] One might want to suggest that there is an *external* 'clock' in the form of the heavenly periods. These cause the length of life to be regular because, on the one hand, celestial periods are regular, and, on the other hand, they affect living things through the succession of hot and cold periods (night and day, the seasons). But it should be noted that such effects depend on what the living thing is: the primary explanatory factor lies in the living thing itself, since otherwise all living things affected in the same way would live for the same length of time. But the celestial cycles still form part the theory of the place of living things, and as such we shall consider them below.

If the life-cycle of different things depends on what they are, then their development and decay need not depend on some clock timing when and how long such processes occur. In considering the development of things, and more generally the temporal ordering of the life of a living thing, the way in which the parts of things grow may play a central role. There is no

12

need to time the growth of things, because growth takes a determinate amount of time, within certain limits. In other words, development takes as long as it does because of the way things grow. The process of development as such takes a certain amount of time. There is no need for a separate clock, since the timing of the processes is inherent in the processes themselves. And when things have grown their parts have a certain resistance to decay which allows them to continue functioning for a certain amount of time. They then stop on account of the errors that accumulate in the course of repairing their tissues. At some point the errors are so many that the living things are no longer viable.

Had Aristotle thought that ageing was best explained by triggered destruction, he would have needed to postulate a separate mechanism to regulate it. As it is, the thesis that ageing is a running out of control corresponds more closely to Aristotle's explanation. For loss of control may not require a special cause, a trigger timed to go off after a certain time. It may just happen anyway. Therefore a biological clock might be redundant; for the way control breaks down can be determined by the organisation of the system, without the breakdown having a trigger. To give an illustration, suppose you build a sand-castle at low tide just below the high water mark. You have to repair it periodically to preserve it against the inroads of the tide. But suppose that after a while you repair it less and less often. The sand-castle will cease to be preserved – it will decay. But it will not decay arbitrarily. The way the water successively wears it out – which parts hold out longest – depends on the design and the quality of sand (if we suppose that the waves behave regularly). In this way, the 'life-expectancy' of the castle depends on its structure. You could have achieved the same end by using a time-bomb. But there is no need for the time-bomb; the castle decays in a regular way anyway. Of course, living beings preserve ('repair') themselves throughout life; but after a certain point this repair is progressively less successful. And the organisation, repair and (to some extent) destabilisation will be due to the organism itself and not to external agencies. The important point is that one can see from this comparison that the course of decay can be determined by structure, and does not require the equivalent of a time-bomb.

Another important and obvious cycle is the daily one of sleeping and waking. The following, extremely brief, look at Aristotle's theory, which is also part of the *Parva Naturalia*, namely *de Somno et Vigilia*,[37] will show us how he explains such regular behaviour. Three aspects can be isolated, which can act as guides to our expectations:

(1) While the living being's structure and activity do not explain the cyclicity of sleeping and waking in general terms, they do explain their occurrence in terms of the activities of the living thing in question: plants

do not sleep because they do not perceive, and sleep is a failure of perception brought on by the process of nutrition.

(2) At another level, waking and sleeping are explained by the fact that capacities are temporally limited for Aristotle.[38] That is to say, a capacity such as perception has an in-built temporal limit. This means that it is not obvious or necessary to ask what this capacity would be like under different circumstances, for example if the cycle of day and night were longer or shorter than it actually is.

(3) A final point worth mentioning is that Aristotle apparently does not think of measuring the temporal limitations of capacities.

Correspondingly for his theory of life-cycles:

(1) We expect no general account of why life is cyclical. Instead parts and activities of living things are integrated into their living cycles.

(2) Life-spans are related to the capacities for change of the living things concerned; but it is assumed that all such capacities are limited.

(3) Finally, there are no measurements.

For Aristotle, the control of living processes is through living forms. If decay can be accounted for in this way, then the way a function exists brings with it the cessation of that function. Although decay involves the failure of function, natural decay is not regarded as bad,[39] and therefore poses no fundamental problem for Aristotle's teleological view of nature; of course, it is a completely different question whether or not premature or violent decay is bad. This is to the good since Aristotle shows no signs of worrying that his account of life and death could cause difficulties for his teleology.[40] He is quite assured that the gradual failure of function that happens in ageing and culminates in death is a natural process. And that would seem to entail that these processes are compatible with the end, that is, 'the best' end of living things. The idea of a life-cycle incorporates the prime of life, when function is most fully achieved, together with the run up to and decline from this phase. Since growth and decline do not incorporate full functioning, they seem to be bad; but if full functioning is possible only through the living being growing in such a way as to reach it, and such a grown body must decay, finally arriving at the failure of all function, then it is hardly surprising that there are no texts in which Aristotle says that death is bad.[41]

Death provides Aristotle with the occasion to quote what is perhaps the only joke in his lectures on natural science:

Further, it belongs to the same study to know the end or what something is for and to know whatever is for that end. Now nature is an end and what something is for. For whenever there is a definite end to a continuous change, that last thing is also what it is for; whence the comical sally in the play 'He has reached the end for which he was born' – for the end should not be just any last thing but the best.[42]

This passage occurs in a discussion of what the student of nature (*physikos*) should study. Its conclusion is that he should study both nature as form and nature as matter. He should neither leave matter out nor give an account restricted to matter.[43] For example, both the study of organs and the study of the functions that these organs fulfil, fall under his remit. Here the joke is that death is an end, but not the one for which the man in question was born. Death does not determine his nature, but it is a last thing because it ends life.[44] Coming to be and growth do not happen for the sake of death but for the form and perfection of the growing thing, towards which the growing thing moves continually.

This passage is the first place in the *Physics* where nature is called an end, and introduces in a very simple way the idea that there are good ends in nature, based on the idea of a life-cycle. If one accepts that the idea of maturity has any application, it is hard to deny that a point is reached at which the living thing has developed its full form, a maximum reached in time. This is in contrast to God's life, which is unvarying. Generation provides a way in which the best end, relatively speaking, can be attained, but the price for this achievement is the other end, death. One of the general conditions for the attainment of such a peak – 'the best end' – is that the maximum is then lost;[45] from the peak, one has to go downhill. Thus once it has reached its peak, the living thing cannot advance further.[46]

One might ask why repair cannot be completely successful in one living thing; in that case, Aristotle would not talk of the coming to be and perishing of a living thing, that is, one going through a life-cycle, but merely of changes in its disposition (*diathesis*).[47] The same individual may not come to be twice. Thus it is the species which is the term in cycles that involve particulars, such as living things, in contrast to the cyclical changes of the elements. And because the cycles involve repetition, they are the way in which permanence, and so necessity, is present in the realm of coming to be. In this way, the cycles of living things provide the possibility of a science with universal application. Conversely, the treatment of life and death, as parts of the cycles, will not be one of particulars as such.[48]

One way of describing living things on the present view would be to call them finite wholes,[49] regarded in light of a norm which they fulfil in time. They have a beginning, a middle and an end, and as such constitute a whole. All the stages in the existence of a living thing are connected to its composite nature of form and matter. As we have already suggested, those features of existence connected to the life-cycle are neither purely formal

15

nor purely material, but due to the composition of both. A stage cannot be isolated and explained by one factor alone. The mature activity of the living thing is central in that it is complete; but at any time in its existence, the living thing is what it is. The form remains throughout the cycle while the matter changes and the amount of matter varies, thus accounting for the cycle.

Aristotle thinks this is the best way that finite things can be arranged, given that they are not God. Cyclic coming to be and passing away enables material things to exist – that is, to exist as nearly permanently as they can as material beings. Cyclicity is not another factor alongside matter and form, but is the way in which living form can exist in matter, as a permanent feature of the world. But although we do not get any justification of the basic status of life-cycles in these forms of life, there is some kind of global justification for the idea in its relation to a form of life not subjected to a life-cycle: finite life is limited by its relation to infinite life. Finite life comes in a wide variety of shapes, sizes and lengths – and only exists as thus specified – and these features may reasonably be expected to be connected to cyclicity. Thus although there may be no timing mechanism in the living thing, the development of its parts and their decay are causally involved in the cycle in question; and which parts are developed differs from one kind of living thing to another. This fact makes it possible for there to be a general fact about living things – all finite things are subject to life-cycles – which is reflected in specific ways in the specific kinds of things. In any case, the explanation that life-cycle will be related primarily not to the universal order responsible for the cycles involving all living things, but to the body and activity of the specific living thing. In some sense, naturally, the living thing remains fundamentally affected by its place in the general order of things; but this general order may not deprive the internal factors of the living thing of their prime role.

In this section I have considered the important notion of a life-cycle, and suggested that the ordering factor for Aristotle cannot lie in some mechanism in the living thing, as it were a time-bomb, but in the structure of the being in question, as something that comes to be. This structure determines in what order it is put together and how it decays, as well as how much time both these processes require. I have also defended the idea that living things nourish themselves as long as they are alive, as a preparation for Aristotle's use of this function in his definition of life. In Chapter 2 I turn to Aristotle's conception of nature. This will lead us to a closer consideration of the natural form and matter of living things, and how it is that the parts of living things can perform functions.

2

Nature

2.1 Nature and nutrition

At one point, Aristotle distinguishes mathematics and astronomy from natural science by saying that the latter studies things that perish.[1] So our topic, his theory of how natural things perish and prevent themselves from perishing, is very close to the core of his understanding of natural science. Let us now turn to his conception of nature (*physis*). Of the two texts in which he offers us full-dress accounts of *physis* (*Metaphysics* V 4, *Physics* II 1), the *Metaphysics* passage is closer to our concerns simply because it makes explicit the importance of growth to Aristotle's conception. Starting from this passage, I shall discuss in this chapter nature and its relation to growth and nutrition. In so doing, I shall also relate nature to the different modes of explanation used by Aristotle (form, matter and the moving cause), more fully than I have done so far.[2] On this basis I shall also take a look at the role of forms as non-intentional ends in nature.

Now, to the different senses of *physis*:

> In one way *physis* is said to be the coming to be of things that grow (*phyomena*) as though one were to pronounce the υ long, in another the primary constituent out of which the growing thing grows. Then again [*physis*] is the source of the primary change being present in each of the things as such that are by nature.[3]

This is the start, but also a resumé, of the account of the uses of *physis* in *Metaphysics* V 4. Nature is here taken squarely as the nature of a thing[4] – furthermore, of a thing that grows. In this section I shall discuss two main points: the identity of the primary constituent, and that of the primary change. The latter is growth, I shall argue, and the former is the matter out of which something grows. This provides the basis for the claim that the formal nature of living things is their nutritive soul, which is the source of their growth. I then discuss this sense of 'nature' further in the next section (§2.2).

The first sense, the coming to be of things that grow,[5] is mentioned explicitly by Aristotle in only one other place,[6] but whether or not he explicitly identifies nature with a change, nature is always concerned with change.[7] What is the significance of the fact that nature is seen as the coming to be of things that grow? It might strike the reader that the status of nature as a process is not at stake, but rather the fact that it is being identified with a process of certain concrete things. While it may or may not be an 'empirical' question which things have natures, it is plausible to think that growth both belongs to the concept of living things and is intimately involved in their existence.

There are at least two changes which might be meant by the phrase 'the coming to be of things that grow'.[8] Most obviously, it might mean their inception or generation, that is, the point at which a new individual comes about. Alternatively, one might think that living things are permanently changing – processes necessary for the continued existence of a living thing ('metabolism') are continually taking place – and it might be this 'coming to be' that is meant. For what takes place in these processes is the change of food into the stuff of the living thing. But before the end of this section, I hope it will be clear that Aristotle regarded these two senses of coming to be as intimately connected; and this is not unreasonable insofar as metabolism is naturally required for the formation of a new living being. So, from the beginning, nature is identified with a process producing growing things.

Another ambiguity, or perhaps simple unclarity, confronts us with the next sense of nature – 'the primary constituent out of which the growing thing grows'. The problem concerns the identity of 'the primary constituent'. Another ambiguity can be removed quite simply, I think, namely that removed later in the chapter (1015a7-10) between 'primary relative to the thing' and 'primary generally speaking'. The first sense, applied to matter, is sometimes known as 'proximate matter', e.g. bronze in a bronze ball, Aristotle's example here; the second refers to the ultimate constituent of some kind of things – as, for example, water may be the matter for all things that can be melted.[9] It seems clear that we are dealing with the matter closest to the thing being produced. For here it is a question of the primary constituent of the growing thing; and it is reasonable to think that the growing thing is taken as such, that is to say, we are not concerned with completely general statements of the form 'm is the matter of X', but with a restriction to cases of growing things. Let us then take it that 'primary' means that matter which in some sense belongs closely to the thing concerned.[10]

The ambiguity I am concerned with in the phrase 'primary constituent out of which things grow' is not in 'primary', but in 'constituent out of which something grows'.[11] The problem is that one would expect a constituent to be part of the thing concerned; yet if the growing thing grows out of something, then we might expect *that* thing not to be a part of it. Hay is not yet part of the horse, and yet one could think that the horse does grow out of the hay. In fact, this difficulty can be resolved by pointing to the fact that the process of conversion from hay to horse goes through different stages; and there may be a stage in which, while the hay is in the horse, it is not yet, in some strict sense, part of the horse. It is, however, what the horse primarily (i.e. that is, immediately) grows out of. How this happens will only become clear through Aristotle's discussion of nutrition.[12]

After matter, one might expect form. But all we get is a reference to 'the source of primary change'. And we will later find reason enough to see in this source the form of the thing concerned, the nutritive soul responsible for its growth. Here, the primary change of which form is the source is not identified further. But consider the massive emphasis on growth in the chapter:

coming to be of things that grow (1014b16);
what they grow out of (1014b17);
the analysis of growth (1014b20-6, discussed below);
comings to be and[13] growth are changes from the form (1015a16).

On this basis, it would be surprising if 'primary change' did not mean growth. Another indication pointing in the same direction may be taken from the concept of matter. For I have claimed that, even if the concept of matter is not clear here, its unclarity lies in the fact that growth and nutrition have not yet been analysed. That is to say, the schematic use of the concept of matter refers us for further clarification to what happens in nutrition. And it is reasonable to think that the first thing a living thing does after its inception is to grow, at least in the simple sense of getting bigger, but also in the sense of developing the parts necessary for its mature life.[14] So growth has a good claim to the title of primary change. Equally, this point begs the question of how it becomes bigger: surely other things must be happening to produce this effect?[15] What we need to understand is growth as a natural process.

A more usual starting point for an account of Aristotle's conception of nature would be 'a principle of change and rest': natural things are those that contain such a principle.[16] But all I have done is to concentrate on a single kind of change, growth, and try to argue for its priority in various respects to locomotion and alteration. Nonetheless, some difficulty is provided for my position by the stipulation that natural things, as such, have the principle of rest in themselves. Like Lewis Carroll's Alice, living things cannot stop growing at will. However, the cessation of growth does depend on our nature, insofar as we have a natural size. And similarly for our natural span of life; our capacity to nourish ourselves reaches a natural end, and we cease living. The other end of life, generation, is exceptional. For such processes are caused from without inasmuch as specimens of the same species act as causes bringing forth the new living thing.[17] Nonetheless, once *physis* has been transmitted to the new living thing, it remains responsible for the growth and nutrition in the new specimen – that at least is clear from *Metaphysics* V 4; and so also for the point when increase of size ceases.[18]

Let us now turn to the brief account of growth with which the account of *physis* in *Metaphysics* V 4 continues:

> Things are said to grow that have increase in size through another thing by contact and either growing together (*sympephykenai*) or growing in addition, such as embryos. Growing together differs from contact, for in the latter nothing is necessary except contact, but in things that have grown together there is one thing in both that makes them grow together instead of touching, and makes them be one by continuity and quantity, but not in respect of quality.[19]

Growth is distinguished from increase in size through mere contact, as for example, more sand may be put on the sand-castle, and so by contact be part of the castle. In things that grow together, there is something which makes them one, alongside the contact. Hay, when it has become horse, does not merely touch the horse. Several points in the chapter so far make one think that this factor is the natural form of something. For it is this that is the source of change in natural things, and change here is principally the change of growth. So the reason that the digested hay is part of the horse is that it has something in common with the horse, and that unifying factor is the form of the horse. In contrast, a sand-castle 'grows' because of an external cause, a child with a spade. There is nothing to distinguish the sand in the castle from that on the beach. Nothing beyond rearrangement need happen to it.

There may be thought to be two aspects to a growing thing. On the one hand, it forms a unity, and on the other it has parts. In growth, the food, when assimilated by the living thing, takes on the form of the living thing.[20] So once assimilation has happened, there is something common to the assimilated food and the body it has been assimilated to, namely the form ('one thing in both').[21] The whole growing body, including the newly assimilated food, is unified by the form.

Anything that grows is not merely a single body because its parts touch one another; it is one body because it contains the principle of its being one continuous body.[22] But different parts of the growing thing have different qualities, that is to say, there are qualities which identify an arm as *such and such a part*, and so too with other parts. If we think of form merely as a shape, then the different parts can be isolated; their mere shape is not what unifies all the parts. (Contrast the way a pentagon's parts are unified.) We therefore need an understanding of form that goes beyond simply the shape of the parts and the whole.[23]

To find a fuller account of form, we can turn first to a work in which, in the nature of things, nutrition plays a central role: *On Generation of Animals*. In the course of his investigation (II 4) Aristotle discusses how the embryo is formed, and asks what the contributions of the female and male are. The answer is: matter and form, respectively. At the end of the chapter he turns to the question of what the form is, and asks how it works on the food provided by the female to produce the parts of the living thing:

> Just as, in the independently existing animal or plant, [the nutritive] soul which uses heat and cold as its instruments (for it is on these that the change brought about by the soul depends, and each thing comes about in a certain proportion (*logos*)), at a later stage produces increase in size out of the nourishment supplied, so in precisely the same way at the very outset, this soul puts together the thing coming about by nature. For, just as the matter from which the being derives its increase in size is identical with that out of which it was originally put together, so too the capacity that produces it is the same as that at the outset, but greater.[24] If then this is the nutritive soul, this is what generates the being, and is the nature of each thing, being present in both plants and all animals. The other parts of the soul are present in some living things, but not in others.[25]

While a full-dress account of nutrition is the subject of the next two chapters, the central message of this passage is crucial to my argument now: the nature of any living thing is its nutritive soul, and this is responsible both for forming the new living being, and for its increase in size when formed. In important respects, however, this message must remain vague here; in particular, the relation of the nutritive soul to other functions requires more elucidation. In *Metaphysics* V 4 we were concerned with growth rather than nutrition; and there is clearly a *prima facie* connection between the two: to grow, something must feed. The connection clearly requires more illumination, but as a preliminary point about nutrition, it is plausible that all things that are alive must feed.[26] And if this process is necessary to the existence of the thing, then we have reason to think that this process is necessary to the nature of the thing.[27]

Nutritive soul is identified first of all as that which produces growth in the existing individual; this is then extended to its function in the case of the initial constitution of the individual. The argument for this identification is by analogy: since the matter in both cases is the same, so must the capacity of the soul be the same.[28] This move, however counterintuitive it may seem, at least provides a rationale for one oddity in the account of growth in *Metaphysics* V 4; namely that the growing together of parts of a living thing is put on the same level as the growth in addition of an embryo. A further problem might be thought to lie in the fact that all living things grow, and so growth does not differentiate them, as their nature should, if it prescribes what they are, but merely provides their generic

21

nature. The answer to this problem is that growth only occurs in a variety of specific ways: it is always the growth of this or that body, and so does delimit what something is, what it can do and what its activities are.

We have already pointed to the two processes that might be referred to by the phrase 'the coming to be of things that grow' (1014b16):[29] the coming to be of a new individual, or the continuous coming to be throughout the life of the individual of its body through metabolism. Our passage from *GA* has made clear that these two processes, in this passage called 'coming about by nature' and 'increase in size', are, if not identical, then at least the same in that two factors, matter and the moving cause, are the same. Thus the thing's nature, its nutritive soul, as we have now learnt, is responsible for its existence both initially and as long as it continues to exist. But it is not alone responsible; rather it has this function by acting on matter.

2.2 Nature as form

So far we have seen that nature is concerned with growth – the way certain things, growing things, come to exist – and embraces both their process of coming to be as well as the matter out of which they immediately arise, as well as – finally and most importantly – the source of this process. We have suggested that this source is their form (1014b17), and that this form is the nutritive soul in any living thing. It is now time to consider more closely the formal nature of things, again starting from *Metaphysics* V 4. Of Aristotle's explanatory factors, form is the one that directs and determines things, including their matter and the process of their coming to be. So now three aspects are to be considered:

(1) the priority of form over matter;
(2) form as end (final cause);
(3) how the end of a process may be responsible for that process.

As we shall see, all three aspects are best understood if one bears in mind that form *qua* nature is nutritive soul. Let us start with the priority of form over matter:

> Then in another way *physis* is said to be the substance (*ousia*) of things that are by nature ... such things that are or come to be by nature, are not yet said to have their nature, although that is present from which they naturally come to be or are [i.e. matter], if they do not have the form and shape. Things made of both of these [form and matter] are natural, such as animals and their parts. *Physis* is the primary matter ..., and the form i.e. the *ousia*, and this is the end of their coming to be.[30]

This discussion of form and its relation to matter starts from the sense of 'nature' we are most familiar with: to say what something is, give its *ousia*, is to describe its nature. Yet it is clear that we are not concerned with all things here; for not all things grow.[31] This confirms the restriction of nature to growing things we found in the last section.

Aristotle first of all defines a sense in which nature in the sense of form is primary: even if the natural matter of something is there, nonetheless the matter is not said to have the nature of the thing concerned, because the matter is not yet in possession of the form. Even if there is nutrition in a horse from which it grows, as long as the matter does not have the form, it is not a horse. Alternatively, if one thinks about growth and asks what it is that is growing, then the answer is: it is the horse that grows, not its food.[32] The food does not determine what grows, rather the form does.

This way of looking at the priority of form over matter with regard to the *physis* of something is based on what we may truly say of something in the course of its coming to be:[33] it first may be said to be a horse, for example when the form *horse* is present. This approach holds out the hope that form will be connected more closely with coming to be.

In both nature and art, form is the controlling factor; but it is also the end of the process controlled. Thus we come to the second of the inter-related topics of this section: form as end (1015a11). The first topic we have discussed, the priority of form over matter, is connected to the status of form as end: the end will take pride of place in a process insofar as the end, and so not the matter, directs the process. And so the form plays the role of the principle of the process.[34] We are now faced with a central problem of Aristotelian natural philosophy: namely how we are to understand his teleology, the end-directedness of nature. The aspect we are confronted with here is particularly difficult, since it may look as though the form is the moving cause ('source of process'); and if we take a thing growing up, which does not yet have its (full) form, how can the form be active? This is the third aspect of natural form that I wish to address in this section. How can something that is not yet there – the form, the end – cause process?[35]

The answer is: it cannot. But the question is, is there somewhere that form might be, when it is not in the offspring? And the answer is: yes, in the progenitor.[36] The answer to this question lies in Aristotle's view of generation: the formal cause that makes the growing thing grow is the form of its father.[37] Thus an external 'shove' sets the process going that produces a new living being with nutritive soul, which will eventually lead to the mature individual. The mature form of the latter does not work backwards through time; it is the form of the growing thing that is active. As soon as the new being is itself an actual being, it nourishes itself and so grows of its own accord until complete. That the process is set off at all, and what the product is, depends on the nature of the initiating 'shove'.[38]

2.2 Nature as form

This is not the place to refine on this rough sketch; but several problems deserve mention. It is very difficult to see how Aristotle could account for the presence of the mature form of the kind in the immature, but actual being. A child is an actual being in that it is human and has a soul, i.e. actuality; in some way it possesses its final form, though not completely. It is essentially on the way to completion. And it is hard to imagine that he has a good answer to the question *where* all the information is, when the form is not actually there. Fortunately, that is a question we do not have to go into. All we need to consider is the presence of the form as the process of nutrition; not how this regulated process is either imparted by the progenitor, or how, prior to adulthood, it has the capacity to develop into the mature form of the thing.[39] We must bear in mind that children, like the aged, are human, and so to be understood in terms of this life-cycle.

By contrast, consider an artefact. One might say that the sand-castle does indeed have a source of change in itself. For it undergoes changes, and the kinds of changes it undergoes are due to itself, that is to say, due to its shape, hardness, wetness, its being made of sand. Thus the way it is affected by its surroundings, and so changes, is due to what it is. But none of this takes us very far towards explaining why it exists. And when we turn to ask why the boy exists, we seem to come to other things, namely his parents. In that sense, of course, the boy is not the cause of himself; that is very clear to Aristotle.[40] But the parents are of the same kind as the boy; and furthermore they have the cause of change in themselves. Like him they feed themselves, grow mature, waste away and die. To say that the boy has a principle of change within himself is not to say that he causes his own existence. It is rather that his activities regulate those changes that contribute to his nutrition and growth.

If one approaches causation, as we tend to,[41] from the idea that a cause is fundamentally something that moves or changes something else, there is a great temptation to try to rescue final causation by seeing it as a type of moving cause. There is, *prima facie*, something to be said for this. When Aristotle says that man generates man, it is clear that what is active in generation is form; and the soul, the living form, is explicitly said to be a moving cause.[42]

But the problem with assimilating ends to moving causes lies in the temporal arrangement of the causes. Moving causes always come before the things they move, whereas ends either come after, or at the same time as the things they explain.[43] It is the living mollusc, as it now is, that is to be explained in terms of its end. And when we consider the boy as a growing thing, the way he grows is to be explained by a form that he does not yet (fully) possess. These considerations would lead one to think that the soul in one respect changes the thing of which it is the soul, but in another way does not. For it determines what a human is throughout his or her life. One can understand this determination partly as nutrition, producing the human form in ever new matter. But alongside this determination of matter, the form can be described or defined. And what is then defined is not, for Aristotle, fitted with a temporal index, identifying one form as the father's form, the other as the son's.[44] Human form is human form, whenever it occurs, and even if the way it exists is by being a moving form: it exists by imposing itself on matter. So we must admit that the formal cause cannot simply be subsumed under the moving cause, although it acts as such in generation.[45]

In the last lines of *Metaphysics* V 4, the priority of natural form, and its identification with nature properly speaking, is made explicit:

> On the basis of what has been said, it follows that primary nature, i.e. nature properly speaking, is the *ousia* of things which have a principle of change in themselves as such. For matter is said to be *physis* by being receptive of *ousia*, and comings to be and growing are [nature] by being changes from it. And *ousia* is the principle of change of natural beings, since it is present in them either in capacity or in actuality.[46]

If we had any doubt that form is meant by the *ousia* of natural things, the contrast between *ousia* and matter in this passage leaves this identification beyond doubt: matter is nature because it takes on form. And if the *ousia* of natural beings is their form, a view we have ample reason from elsewhere in the *Metaphysics* to accept as Aristotle's, then their form is also the principle of change in them. And it is this that is the nature of something, properly speaking, as Aristotle remarks in these concluding lines of *Metaphysics* V 4.

As often, Aristotle tries to establish the systematic relations between the senses of a word. And as usual for him, the way he does this is by relating the entities signified by the word. So here he relates matter and the process of growth to the form of the living thing. How is form to be understood as the principle of change in the light of the importance of growth in the rest of the chapter?

He has already laid stress on the unity of the growing thing (1014b25); and we have suggested that the factor responsible for this unity is form. The food that a growing thing takes in has to be assimilated to the formal nature of that thing – otherwise, how could it serve as food for this thing? And so this process of assimilation has to end in this formal nature – for how otherwise can it be *assimilation*? That is to say, it has to be made *like*, to take on the form of the being concerned.

The source of process in the natural thing (1014b18) is identified with its *ousia*, or form (1014b35 ff, 1015a5). I have already touched briefly on the problem of the effectiveness of a form that is not present. In the final lines of the treatment of *physis*, this point is touched on again. The form is not present in the growing thing, it is only there *in capacity*, insofar as the thing is not complete or mature; the form is in the thing in capacity because the thing has the capacity to develop, namely as something actually nourishing itself.[47] At a certain level, this can simply be understood to mean that the growing thing has the capacity to arrive at maturity; the growing thing has, of itself, the capacity to reach full adulthood (1015a18-19). In other words, the living thing develops of its own accord, requires no further 'shove' from outside for its capacity to grow to be realised.

How can one take the idea that form serves as an end here? A first thought might be that growth ends when growth is complete. Yet maturity is not just an end, it is the principle dictating the road needed to reach it. Saying that living things mature of their own accord means simply that an end is reached in nature, and that the process of reaching that end is ordered in such a way that the end is reached.[48] But even if growth is regular, there may be other ways that this regularity can be explained, for example purely in terms of the material nature of the system. That is to say, the question is why the regularity of growth has to be explained in terms of the final cause, if we are to defend Aristotle.

Consider the phrase: 'principle of change' when applied to growth. Suppose there is a way in which one kind of living thing grows, so that it makes sense to talk of mature specimens. Then, insofar as we are talking about changes *contributing to reaching that mature form*, those changes will not change that form.[49] The changes contribute to, and do not affect the end reached.[50] Living things are necessarily changing,[51] and they change in characteristic ways. The way they change does not itself change: it is, as it were, the regulation of the changes. The same idea is implicit in talk of assimilation as what happens in nutrition: a growing thing imposes its own regularity on things other than itself. And in so doing, the living thing does not change from being what it is; on the contrary, it preserves itself as such. In other words, on this view, the changes involved in nutrition are regulated according to a principle. And that principle is form.

An exception to the rule that a thing only nourishes itself may seem to be given in the case of an embryo or seed.[52] For then a new individual of the same kind (*eidos*) as the parent arises. But it is the same form (*eidos*) that is being nourished; in that sense, nutrition is always of the same kind. This is the reason that Aristotle treats 'growth-in-addition' (i.e. of embryos) and 'growing-together' both as variants on the theme of growth (1014b20-6). We have not, by talking about generation, moved away from the subject of growing, as will become clear when we consider nutritive soul, Aristotle in fact identifies the activities of nutrition, growth and generation. The connection between nutrition and growth is obvious; that between these process and the production of progeny less so. But that is a topic for later consideration (see below §3.3).

2.3 Heuristic and ontological teleology

In the last section I broached the large question of teleology in Aristotle, and suggested that the natural form of living things can serve as their end if the form is understood as regulating the process of growth. I shall now pursue the subject a bit further. This is not the place for a complete survey of teleology in Aristotle, and I shall confine myself to considering those aspects related to organisms, and, more specifically, to those aspects concerned with nutrition. In many ways this is a test case for Aristotle, since on the one hand organisms are central to any consideration of his views, and on the other, if nutrition is not amenable to teleological explanation, then the basis of the existence of living things is not end-directed. And if that is not teleological, then we might have reason to question whether or not living things as such can be end-directed at all.

My approach will be through a concept closely related to Aristotle's theory of nature, namely art (*technê*).[53] Two problems will be considered:

(1) Intentionality in nature: Aristotle uses the model of *technê* to describe living things, and the question is how good such an analogy is: in nature (in the relevant sense) intentions, central to the idea of end-directedness in production, are missing.

(2) The use of the art model, not to make claims about the way things are, but as a model for investigating things. The question immediately arises why one should use a model, if it is not demonstrably applicable on independent grounds.

One way of deciding which of form and matter is primarily nature is to ask when something has its nature; and the answer is when it has the form, and not when it is merely matter. This criterion for the identification of a nature is not enough to distinguish art from nature: sand suitable for a sand-castle is not yet a sand-castle; and no sand-castle is a natural being.[54] The point lies not so much in the difference between growth and making, considered merely as ordered processes: that is the reason that in both cases the relevant matter is not yet the complete thing. Rather, the question is where the controlling factor of the process is located.

The analogy between art and nature is closely related to the last section. There, I asked how it is that the end, although not yet reached in the course of development, can be a cause. In artistic production the 'intention' i.e. the form of the finished product in the mind of the artist, can precede and guide the result; in natural processes all that seems to precede what comes about is the moving cause. In particular, we mentioned the way in which the moving cause in the generation of living things is responsible for the presence of form in progeny; and we left open the question of quite how Aristotle thought this form could be present in capacity in the living thing, in such a way that a mature specimen ensues.

The passage from *GA* II 4 that we discussed above[55] for its identification between nutritive soul and nature is introduced as follows:

> Now the products which come to be by art come to be by means of tools, or rather it would be truer to say that they come to be by means of the change caused by the tools and this change is the activity of the art, for by art we mean the shape of the products which come to be, although [the shape in the sense of the art] is resident elsewhere than in the products themselves. The capacity of the nutritive soul behaves in the same way.[56]

The main thesis here is that change is the vehicle of the order produced, both in nature and in art. Change is not form, it is the way form is reached. In art, the origin of change is situated in something other than the thing produced. The artist produces form in matter through tools. In contrast, the nutritive soul is present in the thing that comes about. In this section, this thesis will be developed a little, and some effort made to show that this is the *only* difference between natural and artificial productions.[57] In the lines following,[58] nutritive soul is identified with the nature of the thing – and so, if for no other reason, is *in* the thing: something's nature must be present in it. And presumably, insofar as something is responsible for its own nutrition, the same must hold for its nutritive soul: it is *in* the living thing. None of this holds for art: *that* is in the craftsman, and acts on something beyond him.[59] Artefacts have no nature, properly speaking, because they are not natural. The similarity between nature and art lies in the fact that, according to Aristotle, the way they both act is through the changes brought about by the tools.[60]

Above, I have attempted to justify talk of form as the regulatory principle of growth by pointing out that in any change not everything can change: there has to be some unchanging feature, and this can be identified with the form, or principle of change. The above text from *GA* II 4 suggests that changes can convey order. And it is not fanciful to see a connection between this transport function of change and the possibility of changes being ordered. A certain arrangement of changes may be able to convey a form, just because changes can be ordered. In this aspect of changes lies the capacity of the craftsman to introduce a certain order to sand to produce a sand-castle; or of humans to assimilate food to themselves.

Now, tools serve ends, so saying that there are tools ('organs') in nature constitutes a claim that there are ends in nature: nature is like art in having ends, but, unlike art, it contains the source of change for reaching those ends within itself.[61] Changes are the path by which ends are reached in both nature and art. So what we have to understand in nature is how changes are regulated in such a way as to bring about ends.

In his discussion of the formation of the embryo in *GA* II 1, Aristotle sets out the principle (734b20-3) that coming to be or generation is to be described as follows. For the coming to be of a human, for example, we need an actual human, under whose influence (*hypo*) a potential human turns into an actual one.[62] In more concrete terms, the semen from the male imports changes into the matter ('menstrual residue') provided by the female. The parts of the living thing are formed by the changes effected through the semen, producing the different parts of the body. The status of these parts will concern us later; but here it is important to note that, while the hot and cold may produce qualities such as hardness or softness (734b31-735a3) they do not produce the proportion (*logos*) between parts which constitutes flesh and bone. Just as with art, it is not the heat and cold that produce the sword, though they may soften and harden the iron. Rather the form, i.e. the principle of the art involved occurs in the *relation between changes* which the instruments perform. Put simply, some specific complex of changes in iron will produce a sword, i.e. will inscribe the form of *sword* in the metal. What transfers the form of a sword into the metal is just this complex of changes.

A more usual approach to form, and one mentioned above, is to ask how it is related to matter. Motion or change and matter are necessarily connected for Aristotle;[63] thus if the formal/final cause is necessarily motive, as the regulation of the changes, then it follows that such regulation must be in matter. No change without matter, no regulated change[64] without matter. Yet matter makes a contribution to substance. At least, if we are unable to make sense of matter contributing to substance, then we will ask why we should bother with matter at all. And it is a feature of some readings of Aristotle to suggest that his gestures towards matter are just that: gestures without the necessary grasp on matter to make them gestures towards something. But one advantage of our concentration on nutrition is that understanding the connection with matter requires no very sophisticated grasp on matter; merely that things that rely on nutrition are material.

Since the form is the end, and in artefacts the form is what guides the producer in his production, it is obviously separate from the artefact. But in the case of natural things, the form never exists in abstraction in this way; so the question arises whether form is really separate in natural things.[65] This attack can be countered by pointing out that parts of living things remain parts only by undergoing change – that is, being nourished.[66] If this is true, and if it is true that the living thing nourishes itself, then we do have a distinction between form and matter in living things. The matter is changing, the form remains. This is connected to the explanation of decay, namely using the body: the body, and not the soul,

29

undergoes decay, as a species of change. But even if the body explains decay, it is the living body that does so, and this implicates the soul too in the explanation.

Teleology can be understood as a thesis about the way things are; but it can also, without implying any thesis about the way things are, be used as an assumption or maxim guiding research.[67] An enquiry into living things, for example, can be guided by the thought that their parts serve ends. And much of Aristotle's research is guided by this thought.[68] Indeed, one might say that his whole enquiry into nature is guided by a teleological model: *technê*. *Technê* is clear to us, we have of necessity insight into its workings. In *JSVM* Aristotle accounts as follows for what he takes to be his greater success in comparison to his predecessors:

> The main reason [Aristotle's predecessors] did not give a good account of [inhalation and exhalation] is that they had no experience of the insides of animals, and they did not assume that nature produces everything for the sake of something. For if they had enquired for the sake of what respiration is present in animals, and had pursued this question with reference to the parts, i.e. lungs and gills, they would have found the cause more readily.[69]

This passage occurs in the extended discussion of his predecessors on respiration in *JSVM*, more particularly in the section on Democritus and Diogenes of Apollonia. Our reactions to it may well be divided – into respect for Aristotle the anatomist, motivating research into organs by the assumption that they perform a function, and reservation about the ascription of such a function to the nature of the thing concerned. And there may be a certain sense in which Aristotle himself might have discovered more, for his accounts of heart, lungs and gills are so profoundly flawed as to be useless.[70]

Furthermore, Aristotle seems here not to be completely categorical about the necessity of grasping the function of natural things for research: he says only that they would have found the explanation more readily, not that in the absence of this assumption they could not find the explanation. And indeed he goes on to discuss Democritus' account of what happens to animals through breathing, and criticises him, along with the other natural philosophers, for not having taken this kind of explanation into account.

Indeed, it might even seem that Aristotle demands, not that one finds causes by postulating ends in nature, but that by postulating ends in nature one finds ends in nature. For this is clearly the cause or explanation that his predecessors would have found, according to Aristotle, had they looked. And indeed his own theory of the lungs and gills is founded on the function they perform in living things. That is to say, rather than being just a thought that we can abandon when we have the result, the assumption that there are ends is necessary if we are to understand what happens in living things. There is no second stage in which Aristotle calls into question the teleological results of his findings, no attempt to rid himself of a cumbersome and possibly misleading methodological assumption.[71] While it is true that simpler levels of organisation than that which exists in living things play a certain role in the explanation of living behaviour, this cannot serve as a level of explanation to which that behaviour can be reduced. While living things are related to the matter they make their own, they have their own natural behaviour, with its own propensities.

Rather than locating his teleology on a reflective level, removed from reality, Aristotle bases his approach to things on the way he thinks they are. And he takes the basic reality of our world to be living form or nutritive process. So the assumption that parts of living bodies serve one another in the performance of this and other functions seems harmless (even if he was in fact disastrously wrong about the functions he attributed to organs). For he is sure that one function is performed naturally: nutrition. And the comparison with artefacts has the added advantage that it connects a sphere in which our insight must be present with one where it is clearly not given. The idea that he was dealing with things in which some parts are necessary[72] because they perform vital functions is one that is obvious in the present field of enquiry: if you wish to decide which are vital organs, vivisection will tell you. This is a path that Aristotle takes, but it is a path that is obvious only if you accept the idea of organs – that is, that parts of bodies serve ends within the vital activity of the whole. It should be noted that this assumption is not one that Aristotle was concerned to falsify, but more simply to see confirmed by his investigation.

Besides these remarks about ends in nature, Aristotle uses comparisons with artefacts to illustrate both the tempering of fire in living things (a kind of brazier: *JSVM* 5 470a8-18) and the structure of the lungs (bellows: *JSVM* 13 474a12-17, 27 480a20-480b7).[73] In doing this he was hardly breaking new ground[74] – witness Empedocles' celebrated comparison of breathing with a water lifter that Aristotle himself quotes (*JSVM* 13). Yet, remarkably, in these comparisons Aristotle is not concerned with the finality of the processes, but in the one case with the mechanism of cooling and heat preservation in an oven; and in the other simply with the structure of the bellows: lungs are like bellows in being double. In fact, considering that he thinks the function of lungs is to cool, it would seem that, for the purposes of this comparison, that function had to be ignored – if we assume that he was aware that bellows are used to produce greater heat in a fire.

For Aristotle, alongside the parallels between art and nature there is a fundamental difference. The parallels lie in the distinction between matter and form, which is another way of saying that some matter is necessary if the end is to be reached. The difference lies in the way the end is reached: the moving cause of an artefact is its producer, and that producer is different from the thing it produces. It is different not merely in the sense of being a distinct thing – the boy is one thing, his sand-castle another; but in being a different kind of thing. And together with this distinction there is another, namely the way producer and produced exist. By this I mean nothing very subtle, only that the one has, as such, its source of change in itself, and the other does not. It might seem that the sand-castle is a bad example of an artefact. After all, there is hardly an art of building sand-castles. And sand-castles are very unsophisticated things in both structure and matter. Yet the point of the illustration lies not so much in the concrete characteristics of the thing; it is rather its the dependence on its maker that is important. That it is arbitrary in structure is not a disadvantage, but points to the dependence of artefacts on the needs and abilities of their makers.

Sand-castles are vestiges left by children; shells are produced by molluscs. Saying that the latter are like the activities of the former in being end-directed leaves one with the question how alike the two really are. For one might think that sand-castle builders are led to select this form rather than that, this amount of such-and-such sand rather than that; and one might describe this process of selection by saying that the builder is sensitive to and directed by the notion of a good sand-castle.

Molluscs, on the other hand, might be thought rather insensitive to the good. Rather, their parts function, that is have a more or less stable disposition to act in certain way, within a certain system.[75] The difference between the two models lies in the idea of sensitivity to the good; and because the good would seem to have a close connection with the idea of choice, the question which model Aristotle is using is of central importance. The functional model seems preferable if there is no choice in growth, for otherwise we would be imputing purposes to things that, although living, have little room for manoeuvre in the way they behave.[76]

Now some thinkers such as Kant see that there is great virtue in the assumption of ends in nature for the purposes of research, particularly into organisms. But if they wish to avoid ascribing teleology to the way things are, a considerable conceptual apparatus is necessary. Above all, as the case of Kant makes clear, *objectivity* must be constituted independently of teleology.[77] If we deny teleology to things themselves, while using it to guide our investigations, then we have to find another way of talking about things themselves. That is to say, if we are to use ends or purposes merely as concepts of reflection, then we must have an independent set of concepts to judge and determine objects.[78] Aristotle can be more economical. For there is no doubt that he did think not only that ends are necessary for our explanations of living things, but also that living things fulfil functions. In other words, ends or forms are quite real for

Aristotle. Indeed, one would be hard put to it to say what for Aristotle, constitutes objectivity – a realm of things independent of the way we think they are – if not formally determined living things.

There is, however, a wide range of passages, particularly in the great treatises on animals, which seem to imply that nature has a choice, in that it is said to do the best thing possible. This is closely allied to the idea that nature produces nothing in vain, that is, natural products serve ends. Yet it is noticeable that such claims often explicitly come as part of reflections about method. That is to say, in reflecting on possible organisations of living things, we are not attributing a choice to nature, but merely coming to understand why it is that the way things are organised could not, under the circumstances, be improved on. There are clearly many questions to answer here about the nature of the constraints on the possibilities reflected on; but the main thing we have to insist on is that nature, even as a producer, has no choice or intention.[79] Unsurprisingly, we understand this position best when we look at it as we would ourselves, when faced with a choice of means to fulfil a task.

In the present section I have pursued the line that nutritive soul presents a functional model of teleology for Aristotle. The end is the form that does not change; this is what distinguishes it from matter and gives it its place in the Aristotelian world. Yet at the same time a soul that feeds needs to feed off something, food, and feeds something, the body it belongs to, so this form or end cannot be considered in isolation from the material parts that contribute to it. As yet we have not done any of this work, and that is the next task: Chapter 3 is about nutritive soul, and Chapter 4 on the body it feeds.

3

Soul

3.1 The project of *de Anima* and its completion in *Parva Naturalia*

In our account of Aristotle's conception of nature (above §§2.1, 2.2), it became clear that investigating things that exist by virtue of such a principle would at least include an investigation into living things. This project of investigation is that principally undertaken in *de Anima* and *Parva Naturalia*.[1] Near the beginning of *de Anima*, he describes the project as follows:

> It seems that acquaintance [with the soul] contributes greatly to all truth, but most of all to nature; for [the soul] is something like a principle of animals. Our aim is to contemplate and be acquainted with its nature and essence, and then its accidents. Some of the latter seem to be proper attributes of the soul itself, others are also present in animals because of the soul.[2]

Later in this introductory chapter, Aristotle goes on to discuss the affections of the soul, and points out that they seem, for the most part, connected to the body – generally speaking, they are linked to perception.[3] In the famous phrase, attributes of the soul are 'enmattered definitions'.[4] Whereas the dialectician would define anger as a desire for revenge, the natural scientist will have to say something about the body; the one gives merely the form and its definition, the other, the matter. Thus the natural philosopher must discuss both form and matter: 'the definition of the thing is such and such, and it is necessary for this definition to be in such and such matter.' Natural science must deal with all four of Aristotle's causes.[5] The soul is the form, the moving cause and what the body of a living thing is for. That is to say, the soul can serve as all of Aristotle's causes except matter. The student of nature should not restrict himself to matter, nor should he ignore matter when studying the soul and living things;[6] but the form has priority, since it explains the matter, whereas matter does not explain the form.[7] Saying that the natural philosopher must talk about both soul and body is not informative, insofar as these concepts are not clear. The purpose of this chapter is to show how soul is defined, and how this definition locates it in a body (matter, in some sense). This will allow

34

us to move on to the investigation of living body in the next chapter. This programme exhibits both aspects of form, as outlined in the last chapter. On the one hand, it guides the investigation; on the other, it exists in the things investigated.[8]

In my discussion of Gill and Freudenthal, I suggested that neither gave a satisfactory account of the role played by matter in Aristotle's psychology. To answer them more fully I shall have to give an account of matter in living things, and I do this in the next chapter.[9] Aristotle thinks that living things, as well as being wholes, are composed of organs ('anhomoiomerous parts'), and organs are composed of uniform stuff ('homoiomerous parts'): the difference between these two is that the latter when divided into arbitrarily small portions is still the same stuff, while an organ such as a hand obviously is not. In order to understand this hierarchy of matter, we will have to understand Aristotle's concept of 'mixture' (*mixis*). Only when we have done that will we be in a position to give anything approaching a complete account of soul's presence in body. This will also have to include some account of the relation between the concepts of 'body' and 'matter'. Not all matter is in living things, and the question arises of what difference its presence there makes.

So the project of *de Anima* includes not merely the soul, but also those entities – attributes, changes, activities – which accrue to living things because of the presence of soul. We shall now put the case for thinking that this group of characteristics will include actions involving body and soul, which form the topic of *Parva Naturalia*. That is to say, *Parva Naturalia* is psychology in Aristotle's sense of the term:[10] the things already said about the soul in the *de Anima* are assumed.[11] As has often been pointed out, *Parva Naturalia* (*Small Treatises on Nature*), is an uninformative title for the work. It deserves a better title, as well as one closer to the way Aristotle himself thought of it. There are two good candidates, both of which can be derived from the first chapter of *de Sensu et Sensibilibus*, which serves as an introduction to *Parva Naturalia*: *idiai kai koinai praxeis* (*Peculiar and Common Actions* sc. of animals and things possessing life) and *koina tês psychês kai tou sômatos* (*Things Common to Body and Soul*).[12] The first line of *de Sensu et Sensibilibus* (436a1) implies that soul 'as such' had already been treated, and now animals and things with life can be discussed – more exactly, what their peculiar and common actions are. It is these actions that are common to body and soul. While both topics separate the *Parva Naturalia* from the *de Anima* to some degree, neither does so cleanly. As we have seen, the *de Anima* also wants to deal with those things that belong to animals in virtue of the soul: and that would include things common to body and soul: for no full account of the senses is possible without material explanation.[13] It also deals with the common actions of living things since the soul defines their most general capacities (nutrition, generation, perception). Thus neither of these phrases makes clear exactly what the distinction between *de Anima* and *Parva Naturalia* is. However one can say that *de Anima* is, unsurprisingly, about the soul as such: that is, all soul, including that which is separable (so-called 'productive intellect' *nous poêtikos*). And *Parva Naturalia* is about animals and 'things with life', especially those actions of living beings that result from their possession of soul, which can then be discussed well after all the capacities of the soul have been severally discussed in *de Anima*. *Parva Naturalia* can then discuss the varieties in which these capacities, and just these capacities – we learn of no new ones – actually exist, that is to say the variety of ways these functions are performed. Much of this work involves fairly detailed physiology. The physiology, however, is not itself the subject, except as it is related to the soul.[14] One might say that *Parva Naturalia* works with the distinctions set up in *de Anima* about the soul, to give a concrete but general picture of living things' actions.[15]

De Sensu et Sensibilibus 1[16] gives a brief survey of the work to come (rather than a table of contents). As we have just seen, the aim of the *Parva Naturalia* is to investigate the common or peculiar 'actions' of living things. The things that are most important are obviously common to body and soul,[17] and he gives examples such as sensation, memory, appetite, and pleasure and pain, which are present in almost all animals. Some of these features are common to everything living, others merely to some animals. He continues:

> The most important of the common actions are in fact 'conjunctions', [i.e. pairs], four in number, such as waking and sleeping, youth and old age, breathing in and out, and life and death.[18]

The 'actions' listed are very heterogeneous; apparently, all that connects them as a group is that they are important, common actions of living things. Some are clearly essential to some living beings, others seem of less importance, or are less generally found, for example, plants do not sleep. Thus coming to be and death are common to all living things.[19] Some of these actions, like perception, serve functions; others, like ageing, do not appear immediately to have a function.[20]

Does *Parva Naturalia* deal with *everything* that is common to both body and soul? It does deal with both nutritive and perceptive soul and their organs. These are certainly the main actions involving body and soul.[21] As to the question of what it means for something to be common to body and soul, we can defer our answer until we consider the individual functions; Aristotle at the start of *Parva Naturalia* merely says that these actions are 'not unclearly' common to body and soul since they either occur with or through perception. And that perception occurs to the soul through the body[22] is clear both on theoretical grounds and independently. In animals, life and death are linked to perception as follows, one may reasonably think: as long as they live they perceive, and death is a lack of perception. Longevity is more difficult to connect to perception, if the link is not simply how long the animal can preserve its capacity to perceive.

The use of 'actions' (*praxeis*) to describe the topic of *Parva Naturalia* offers certain difficulties. First, intentions and perhaps even rational deliberation might be thought to be necessarily involved.[23] But this point will not help us much with the science of living things in general. Obviously, in some contexts action is unique to humans.[24] However, what is restricted to humans is *rational* action; as long as one is clear about not being restricted to rational action, it seems harmless to say that living beings perform actions; after all, they do things. It is quite clear from the way the study of animals is approached in *de Partibus Animalium* that their actions are central to their study:[25] parts of animals are explained by the actions they are able to perform.[26] Thus the first problem can be solved. 'Action' can apply to living beings other than humans.[27] Even though all of the processes in *Parva Naturalia* are things living beings do, and are related to the end-directed constitution of such beings, there is an added element of flexibility to be noted. The binding of actions in *de Partibus Animalium* to ends in a narrow sense (if x is an action, x serves an end) seems to be broken in *Parva Naturalia*; actions such as dreaming or ageing, not to mention dying, while they are things that only living beings do, do not obviously directly serve ends, unlike, say, perception. So here we have another aspect in which *Parva Naturalia* differs from *de Partibus Animalium*: non-functional actions can be made comprehensible within the whole living being.

The second difficulty concerns the usual understanding of an action (*praxis*) which is incompatible with, for example, sleeping and waking being actions. According to the usual understanding, action understood as an activity (*energeia*) is to be contrasted to change in the following way.[28] As an activity, action has its end within itself. A change, however, itself is incomplete: it has to reach a stage at which it itself ceases, for it to be complete. Walking to Thebes, a change, is complete when you arrive and are no longer walking. Seeing, an action, is complete in itself; it does not reach some end outside itself. You do not have to reach some point in the process before you can say that the process of seeing is complete. As soon as you see, the act of sight as such is complete. Of the topics discussed in *de Longitudine et Brevitate Vitae* and *de Juventute, Senectute, Vita et Morte*, on this reading of action, only life seems to be one.[29]

The problem of the subject of *Parva Naturalia* can be made clearer if we consider what the alternatives would be to calling perception, memory, sleep and waking, dreaming, the length and stages of life 'the actions' of living things. Two possibilities spring to mind: the activities (*energeiai*) or changes (*kinêseis*) of living things. We have seen that activities would be too narrow to encompass the processes Aristotle has in mind; and as for changes,[30] while some processes such as growth are paradigmatically categorical (quantitative), and so neatly comprehensible as a change, *Parva Naturalia* is not actually about such processes, but about complicated organic changes such as dreaming or changes in the course of a whole life. Change in the narrow sense of course plays an important role in such complicated processes; but that does not justify calling these processes 'changes'. They are actions that only the living can perform, and are necessarily related to the life of living things. This relation has two aspects. On the one hand the life of the living thing will be crucial to explaining the actions of the living things. But one hope the reader may have is not only that the actions will be explained, but that they will explain the life of the living thing.

After this general introduction to *de Anima* and *Parva Naturalia*, we must consider the two treatises that will occupy us for much of the time – *de Longitudine et Brevitate Vitae (LBV)* and *de Juventute, Senectute, Vita et Morte (JSVM)*. The title *On the Length and Shortness of Life* has at least some ancient authority: the catalogue of Aristotle's works ascribed to Ptolemy Chennos contains *LBV* under the title *On the Length and Shortness of Life, 1 Book (peri makrobiotêtos kai brachubiotêtos* a').[31] The first four chapters approach the task of applying work from elsewhere in Aristotle's natural science to the problem of longevity. The final two chapters offer solutions to the question of what makes things long-lived or short-lived, without any detailed physiology. In itself, *de Longitudine et Brevitate Vitae* is a coherent investigation of its topic. The cause identified for the length of life is the mixture of heat and wetness in living bodies, but this must be seen within a wider framework for these qualities are fundamental to Aristotle's explanation of living things in general. They are closely connected to the function of self-preservation. The aim of *de Longitudine et Brevitate Vitae* is clear: Aristotle wants to show why some living things are long-lived and others are not.[32] That is, he wishes to identify the causes of longevity and short lives.

Our second treatise has often been thought to form several works. Bekker divides his pages 467b10-480b30 into two treatises: *de Juventute et Senectute* (467b10-470b5), and *de Respiratione* (470b6-480b30).[33] However, there are good reasons to consider these pages as a single treatise on youth, old age, life and death, which might be called *de Juventute, Senectute, Vita et Morte*. No manuscript marks the *de Respiratione* off as a separate treatise.[34] *de Juventute, Senectute, Vita et Morte* is announced at the start of chapter 1 as being about youth, old age, life and death.[35] Respiration (*anapnoê*) is introduced because 'living and not living happen to some animals because of this', and so it must be investigated.[36] It forms no separate subject. In itself, this might not tell against there being a treatise on it within the framework of *Parva Naturalia*. Some topics are clearly subordinate to others, for example, dreams and their interpretation to sleep, and this is reflected in the relevant treatises in *Parva Naturalia*: *de Insomniis* and *de Divinatione per Somnum* are subordinate to *de Somno et Vigilia*. The decisive factor that speaks against a subordinate treatise on respiration is that respiration is a cause to be invoked in the treatment of youth, old age, life and death.

Further reasons for regarding Bekker's pages 467b10-480b30 as a unity are to be found in the way the subjects are treated. In the so-called *de Juventute et Senectute*, i.e. chapters 1-6 of the work called *de Juventute, Senectute, Vita et Morte* here, little is said about youth and old age. *JSVM* 5 does lay the foundation for the later account by distinguishing the natural cooling that happens in old age from the unnatural cooling in violent death. But that is not an account of *youth* and old age. Only in the so-called *de Respiratione* 17 and 18, i.e. *JSVM* 23-4, do we find the explanations of youth, old age, life and death all together, as though they formed part of a coherent whole. That they should do so is clear: the explanation of the stages of life forms part of the explanation of life as a whole. 470b1-5 makes clear that the so-called *de Respiratione*, i.e. *JSVM* 7-27, is purely about the cooling necessary for the preservation of heat and hence life, continuing the treatment of this subject begun in Chapter 6. That is to say, it is dependent on *JSVM* 1-5, and so forms part of the account of youth, old age, life and death.[37]

The alternative to calling this text *de Juventute, Senectute, Vita et Morte* would be to call it *de Vita et Morte* (*On Life and Death*). This is how Ptolemy Chennos refers to the whole text.[38] So why do we have to mention youth and old age alongside life and death? Are they not anyway merely adjuncts of the central phenomena – life and death? The first reason for preferring the more inclusive title is that both the reference at the end of *LBV* (6 467b6-9) and that at the start of *JSVM* (1 467b10) mention both youth and old age; and we have no reason to disregard these remarks. There is, moreover, in the way in which Aristotle thinks of mortality, another reason for considering youth and old age as integral to the enquiry. His concept is not of a constant, unvarying life. For Aristotle, living things develop and decay naturally. This is why youth and old age are so important: as suggested in our introduction, the fundamental idea is that of a *life-cycle*. *JSVM* gives an account of life-cycles in life through living things' capacity for self-preservation, i.e. nutrition, and its failure. As we have seen, *Parva Naturalia* is about the actions of living things, involving both body and soul; and *JSVM* is not to be considered as a treatise merely on the physiology of the heart and related organs:[39] primarily it is about vital functions, in particular nutrition. At the end of the *JSVM*, Aristotle says he has dealt with life and death and those things related to this enquiry.[40] The most important things related to life and death are best taken to be the organs whose functioning is necessary to life, and the failure of which causes death – but we must also bear in mind the stages of life.

After announcing its topics, *JSVM* starts with a reference back to the account of the soul in the *de Anima*.[41] This underlines the connection, which we have already emphasised, between the topic of *JSVM* and the soul. If the *de Anima* investigates the soul, there is another enquiry precisely into the soul as it occurs in the body. Another way of describing this subject might be to say that it is the way in which the body has a capacity for vital function. This has been prepared already in the *de Anima*, and it is to this passage in the *de Anima* that we are referred at the beginning of *JSVM*. In the *de Anima*, we learned that it is a good assumption that soul is neither body nor without it; for the soul is 'something of the body', and what is more, it is in a body of a certain kind: '[Soul] is something of body and because of this it is present in a body – and in such a body.'[42] Aristotle then complains that his predecessors fitted the soul into the body without saying what kind of body it was in. Actuality, he says, comes to be in appropriate matter. That is to say, the soul will be located in matter, i.e. in organs suitable to carry out the functions of the soul. Thus one task that *JSVM* can tackle is to show where the seat of the soul is. The reason for seeing a discontinuity between the *Parva Naturalia* and the *de Anima* is that there is thought to be some incongruity between seeing the soul as an actuality, and as seeing it in the body.[43] However, a proper understanding of the soul as the actuality of the body, more especially in nourishing itself, disposes of this problem. There are anyway enough indications in *de Anima* that the functions of the soul in sanguineous animals are closely connected with the region around the heart for it not to be surprising that a detailed treatment of the body and soul will say more about this.[44] But before locating the seat of the soul, we must first discuss what it is, generally speaking, and then its most fundamental part, nutritive soul.

3.2 The first actuality of a natural body with organs

Soul and the way it exists in matter are, as we saw in the last section (§3.1), the subject of *de Anima* and *Parva Naturalia*. Soul is form, and I have claimed (§2.3) not only that form exists – so it makes sense to ask how it exists – but also that it enjoys a special place in an Aristotelian enquiry as a guide to what to look for in living things, namely tools or organs with the capacity to fulfil or serve the form. The purpose of this section is to present the view of Aristotelian soul as form or actuality (*entelecheia*).[45] In a way, this is completely uncontroversial, if not very informative. Consider the three formulations of Aristotle's definition of the soul (*de Anima* II 1):

(1) So it is necessary that the soul is substance (*ousia*) in the sense of form of a natural body having life in capacity (412a19-21).

(2) Hence soul is the first actuality of a natural body having life in capacity (412a27-8).

(3) So if one must say something common to all soul, it would be the first actuality of a natural body with organs (412b4-6).

At a merely verbal level, only 'natural body' is common to all these formulations; to see how the other parts fit together, let us look at how they are reached. Aristotle's procedure in *de Anima* II 1 starts from the assumption that soul is substance; after taking the category of substance, and dividing substances (412a6-9) into form, body, more especially natural bodies (412a11-12), and composites of form and body, he can then isolate the bodies soul can be in. He continues the division by observing that some natural bodies have life and some do not. And as to life:

we call life self-nutrition and growth and wasting (14-15).

In this way we arrive at 'natural body' in all three formulations, justifying our approach to soul through the concept of nature (§§2.2, 3.1); and the 'definition' of life just given supports the connection between nature and nutrition that guided us through our discussion of nature (§2.1). It was also there that we saw that such a life as represented by nutrition can only be conceived of as in a body. Thus the three elements in the formulation, life, nature and body, are comprehensibly connected.

However, this is not a real definition of life, in that no causes are given, merely minimal identifying characteristics:[46] Aristotle starts from the way we talk about life. A further point is that he emphasises the cyclical character of life. The whole enquiry begins from this quasi-definition and then attempts to show what factors are at work in realising this end – the one end achieved by living things as such. The enquiry then closes with the definitions of the life-cycle, using the nutritive soul, in *JSVM* 24. This text has never, as far as I know, been mentioned in discussions of *de Anima* II 1. It is crucial to seeing what the limitations of that text are.

Soul is form, since of the three ways of being a substance, form, matter and composite, soul is not the last two, for a reason we cannot go into here.[47] The differences between the formulations are twofold. First, the form of the natural body in (1) is replaced by its actuality in (2) and (3). Secondly, the possession of life in capacity in (1) and (2), i.e. the capacity for life, is replaced by the possession of organs in (3).[48] These organs are exemplified (b1-4) by the organs a plant has to reproduce and acquire nutrition. The connection between having organs and having a capacity for life requires no comment: to say that living things can live is to say, at least, that they are equipped to do so. But the substitution underlines an

41

aspect I have already been at pains to stress: right from the start the presence of organs is inscribed into Aristotle's understanding of the soul. And the thought suggests itself that further investigation of soul will have as its correlate, further investigation of the body serving it. This fits in again with the results of the last section, where we saw that the *Parva Naturalia* completes *de Anima* by considering actions common to body and soul, as well as our view of form as guiding an Aristotelian enquiry.[49]

The second difference between the (2) and (3) concerns the way in which actuality is qualified: as *first* actuality. And this is where my reading becomes controversial. At 412a9-10, where form is equated with actuality, we have been prepared for some qualification of actuality by the observation that form is actuality in two ways, first as knowledge and secondly as contemplation (sc. of the knowledge). This line of thought is carried further at 412a22-7, where the assertion of an ambiguity in actuality (*entelecheia*) is repeated and we are told that soul is clearly like knowledge, since both sleep and waking depend on the presence of the soul for their existence: waking is analogous to contemplation, and sleep is analogous to having knowledge and not activating it (*energein*). This last assertion makes it seem obvious that when Aristotle says that the soul is an actuality, he means that it is one as a capacity.[50] And that is the way soul in Aristotle is generally understood.

There are, however, a variety of problems with this reading. To start with, it is not clear that saying that actuality is like knowledge in not being exercised, means that such an actuality is merely a capacity, and not also in a strict sense actuality.[51] The same thought is suggested by the following consideration. Knowledge is said to be prior in generation to contemplation. What this means is that in the process of learning one first of all acquires knowledge, and then is able to contemplate it.[52] But if such coming to be is to be complete, whether of knowledge or of a living being, then something actual must be reached. This is required by Aristotle's view of change: someone has turned pale when he is actually pale. An ordering, such as is suggested by the phrase 'first actuality', can be understood very simply with reference to a process of coming to be: the first actuality reached in the process of generation is the soul of the (new) living being. Coming to be has taken place when the new living thing participates in living activity. This is a very attractive reading of 'first actuality'[53] in that it does not require the reduction of the soul to a mere disposition. The first actuality is thus that which marks the first emergence of an independent actuality in the generation of a living thing. Something actually exists when its end is there, and its end is its function.[54] To ask when something exists fits neatly with the emphasis on the cyclical character of life (412a14-15), but also with the idea of self-nutrition mentioned in the same place: a new living thing exists when it nourishes itself.

Now, it is a common way of talking in *de Anima* to say that there are capacities of the soul – this applies to capacities for nutrition, perception, thought.[55] So the reader of Aristotle who thinks the soul is a capacity has a wide choice of activities (*energeiai*) or actualities for which the soul is a capacity or disposition.[56] Being for a living being is living, as Aristotle says.[57] Alongside the 'definition' of life that we have already met in II 1 – self-nutrition, growth and wasting, all of which are the business of the nutritive soul – a fuller list includes locomotion, perception and thought.[58] A capacity for any one of these might count as sufficient for the presence of life.

For the friend of actuality, the choice is a great deal more restricted; indeed, one could be forgiven for thinking that there is no choice at all. Only nutrition seems to come anywhere near being an actuality that exists as soon as the soul does; and that would seem to be a minimal condition for using an actuality in the definition of soul.[59] And that was precisely our thesis in Chapter 2 above. Further confirmation is also provided by the definitions in *JSVM* 24 (see below Chapter 5).

Living beings, i.e. ones with souls, essentially nourish themselves. The most basic living things, plants, simply possess this soul,[60] specifically differentiated, of course, in the different kinds of plants: nutritive soul *tout court* is present in no living thing. Similarly, in living things where this soul sustains other capacities and their activities, the nutritive soul is present in a differentiated form. The basic soul provides the capacity to the further kinds: in the usual terminology, they are nested – the more basic soul is present *in capacity* in the more developed one. Thus animals that perceive also have the capacity to nourish themselves; humans that can think, can both perceive and nourish themselves. But this nesting works only because the basic soul is actual: with no actual nutrition, possession of the nutritive soul would be little help. The mention of sleep in the passage under discussion indeed provides further confirmation that when asked to name an actuality present in sleeping animals, Aristotle's choice is nutrition. For much of his account of sleep is concerned with nutrition.[61] And we can explain why this actuality (nutrition) provides the capacity for other activities. For the organs that enable animals to perceive and move are products of nutrition.

There is one perhaps decisive argument against this view, namely that the account in *de Anima* II 1 is not meant to fit any one kind or part of soul; for it would not then fit the others, and so not be the most common account of soul, as suggested in the third formulation. For this reason Johannes Hübner, who has argued for this view of the soul in a recent paper, does not suggest that nutritive soul is meant in II 1, but *organic self-maintenance*.[62] By this he means the basic organic actuality (and not merely capacity) which is expressed in self-nutrition, growth and wasting. The problems with this suggestion are twofold. On the one hand, it is not clear

43

that it escapes the difficulties associated with reading the first actuality in II 1 as nutrition; Hübner sees this difficulty, and wishes to avoid it by showing how other vital activities are accommodated within self-maintenance. His moves here are convincing, and we will go into them below. But of course if one can claim this advantage for organic self-main-tenance, one will also be able to claim it for nutritive soul, if this is taken to include both capacity and actuality of nutrition, as is reasonable. To speak of a capacity of nutrition is commonplace in *de Anima*;[63] and that nutritive soul must at least sometimes include its actuality is proved by the fact that some actual living things, plants, merely have this soul. If nutrition remained for ever an unrealised capacity, it would keep nothing alive. None of this implies that what is defined in *de Anima* II 1 is the nutritive soul; merely that one has to give the answer 'nutrition' to the question what the first actuality is.

The second difficulty is that organic self-maintenance would seem to be a kind of soul, insofar as this is to be understood as vital actuality, not otherwise mentioned by Aristotle, and this goes against the stricture that there is no kind or part of soul besides those enumerated by Aristotle which would serve as the thing defined by a general account of soul. This is said in the context of the comparison of the kinds of soul and the difficulty of defining them with the kinds of (geometrical) figures and the difficulty of defining geometrical figure in general: the point is that there is no figure besides triangle, rectangle etc. which encompasses them all. So there is no soul beyond nutritive, perceptive, etc. encompassing them all. And yet that seems to be the way organic self-maintenance has a claim to encompass all souls: identical with none, but basic to all.

Hübner's strategy is to say that organic self-maintenance is not identi-cal with nutritive soul, because the latter involves separate tasks, whereas organic self-maintenance represents the end or achievement of these tasks. The converse of this is that organic self-maintenance is an actuality, whereas the processes involved in the different tasks of the soul are not, but are merely capacities. In support of this view of the different parts of the soul as capacities involved in processes, and not the activity arising from these processes, Hübner points to the capacities identified with the nutritive soul.[64] While it is certainly true that nutritive soul is a capacity, as we have already said, it is not so clear that it is *never* an actuality. Indeed, the case of plants requires that it is an actuality; if it were not, there would be no actuality in plants. In other things, assuming that if anything is to perform the task of nourishing the being,[65] this will be nutritive soul, then nutritive soul will also encompass the end or achieve-ment of nutrition; even if the things nourished, the body in all its parts, serve other ends, the end of nutrition will be arrived at by nutritive soul. So if nutritive soul has to be understood, like other forms of soul, both as capacity and as actuality, organic self-maintenance can be understood simply as actual nutrition.

As to the other problem: how is it that actual nutrition encompasses all the other parts or kinds of soul? The basic thought is simple: nutrition is responsible for and produces the organs involved in vital actualities. Since it is responsible for the growth of the whole body, it is, trivially, responsible for perceptive parts or organs as well. In each case, nutrition or self-maintenance will be carried out in such a way that the whole body is nourished or preserved. What the living thing is, is inscribed into its actual nutrition – this applies equally to the different kinds of plants and the different kinds of animal. Thus this basic actuality encompasses the capacities (organs) for all the other vital actualities. But saying that soul is this actuality does not in fact define or specify all kinds of living thing, rather it is an abstract definition, such as in Euclidean geometry one might want to say of 'plane figure' that it must surround a two-dimensional space.[66]

Saying that the soul is a first actuality of a natural body with organs may be uninformative for a very simple reason: we do not know what actuality is, and Aristotle nowhere defines it; we must learn how to use the distinction between capacity and activity or actuality by understanding examples. The relation between activity and capacity is analogous in the different cases, in particular a capacity for change and the change itself, and the matter for a substance and the substance itself.[67] The Greek word translated by 'actuality', *entelecheia*, an invention of Aristotle's, may be taken to mean that the entity concerned has (*echein*) its end (*telos*), and this in turn implies that such an entity has no further end. Put positively, that would be to say that the soul of a living thing is that aspect of it which is complete; put negatively, it is that aspect which serves no end beyond itself. The concept of actuality was introduced in *de Anima* II 1 instead of form, and so in contrast with matter and capacity. And the reading of actuality here also offers a contrast with capacity; for a capacity is always related to something else. This means that such an entity as an actuality can serve as a containing system, in particular for capacities; for capacities are defined by their relation to things beyond themselves: they are capacities for something.[68] Thus all the capacities of the living thing will serve the one end. In other words, all the vital functions are integrated in the system formed by the body. A living thing has the principle of change within itself since it is able to change (grow, move, perceive) because of capacities it possesses; and the actuality of these capacities for change does not have an end above and beyond the living thing. Order in living things is preserved by change; these changes are ordered in such a way that they have their end in just such an ordering of changes as themselves – they replicate themselves.

Again the problem of the other kinds of soul presents itself here; for it is essential for animals to be able to perceive. And so all the capacities of the animal will be directed to this end. This, however, is a repetition of the point about whether nutrition can encompass all vital activities: perception is preserved by nutrition. Yet the question of which is the end is a serious one: is everything directed at perception or at nutrition? That nutrition undoubtedly serves as a capacity (414b28-32) might indicate that the end must be perception. And that some capacities, notably the capacity of humans to think, are not concerned with mere living ('survival') but living well, should tell us that the real end is the activity supported by nutrition, and not the nutrition itself.[69]

Throughout, I have been talking as though the distinction between actuality and capacity (i.e., as will be remembered from *de Anima* II 1, the distinction between form and matter) is clearly understood and unchallenged. That is not the case. In the case of living things the form-matter distinction is most contentious, as has often been stressed.[70] The problem with living beings is that their matter never exists without them, unlike the matter of artefacts, which are the first examples we use to understand the form-matter distinction. In the case of an axe, the metal is there before it is formed. But this does not apply to the flesh and tissue of a living being.

This problem can be solved, I think. One can ask what it is about a living being that gives it its capacity to be a living being. Aristotle himself does this, and, as an example, warns us against thinking of a corpse as something with a capacity to live.[71] For only something alive can live.[72] Yet there is still a difference between that which has the capacity and the activity itself. Thus the living body has the capacity to be alive, since it is structured in such and such a way; and its being actually alive is the fulfilment of these functions.[73] The relevant distinction is not that between iron that can be, but is not yet, an axe, and iron that is actually an axe, but rather between the capacity of the axe to cut and the cutting of the axe. In the case of the hand, we must distinguish between the hand and its action, for example grasping. That is the relevant distinction between capacity and actuality in living things. In artefacts, the capacity to cut must be distinguished from the matter; in living things, this distinction can only be made in a much more abstract manner. What this means is something very basic to any analogy: keep your eye on the similarity, forget the differences. Thus the pre-existence of metal to its being made into an axe is not relevant, but the axe's capacity for the activity of cutting is.[74] The contrast is that between the matter of living things and what it *does*.[75] The contrast between form and matter, as it appears here, is best understood in terms of the contrast between capacity and activity. Thus the matter has a capacity, e.g. mouth, stomach and so on, to digest food, and the activity associated with this capacity is the digestion ('concoction') of food itself.[76] In other words, the capacity is something other than the actuality, while at the same time being determined by the activity, since that is what it is a capacity for.[77] Another point that Aristotle makes about capacity and

46

activity is that they cannot be defined. All he says is that capacity and activity only become clear when considered together; there is no account of them to be had outside their relation in individual cases. He describes this relation generally thus: being in capacity is not being in activity.[78] This may seem an unsatisfactory remark as a way of explaining being in activity, and this passage in *Metaphysics* IX 6 is directed to someone asking for a fuller account. But we get no further account because we have to learn to use the distinction simply by means of examples.

It is common among interpreters to see the use of the phrase 'first actuality' in *de Anima* II 1 as complemented by the distinctions made in II 5 concerning capacity and actuality.[79] In that chapter it is explained to what extent the alteration that occurs in perception is like other altera-tions. The distinction is that in other alterations one contrary is destroyed when one comes about, such as Socrates' darkness when he turns pale. In other cases, there is no change destroying a contrary, merely the expres-sion of a disposition: a builder when building remains a builder, expressing this disposition in his activity. Perception is then said to be of the latter kind of alteration. The traditional, if un-Aristotelian, terminology is help-ful here.[80] First capacity is being able to acquire the capacity to perceive. Second capacity is being able to perceive; this is at the same time first actuality. And, finally, second actuality is using the capacity to perceive, actual perception if you will. As well as using the distinction between having knowledge and contemplating it,[81] Aristotle also uses a comparison with the capacity to be a general, which all men possess. The distinction is then between the capacity of a child to be a general (first capacity), and that of a man of age to be so (second capacity, first actuality). The child must first change in order to become a general, i.e. grow up; whereas the mature man does not need to change (once he has been elected to the post), but merely exercises his capacity (second actuality).[82]

Aristotle himself does not refer to II 1 in II 5, and the neat scheme – first capacity, second capacity, first actuality, second actuality – is not due to him. Even if this point is the same in both II 1 and II 5, and it is mentioned quite generally in *de Anima* II 1 (412a22-7), the doctrine is only developed with reference to perceptive soul (II 5), i.e. after the treatment of nutritive soul in II 4. There seems to be no room for the contrast between first and second actuality for nutrition, since living things are always nourishing themselves. This is true only up to a point; for we can distin-guish between the capacity for nutrition and actual nutrition, even if no living thing merely has the capacity, and only living things have the capacity for nutrition. Actuality and capacity are different aspects of the living thing, as we shall see when we turn to the definition of nutrition (see below §3.3). For the definition of the soul, the point is that, while nutritive capacity and actuality may be different, actuality must be there for the capacity to be there. As we have pointed out, the central question is what happens in sleep, and there is no doubt Aristotle thought that nutrition also happened in sleep. A text which says nutrition is constantly actual as long as the capacity for it is there, is II 2 413a28-31:[83]

> For plants do not grow upwards without growing downwards; they grow in both directions equally, in fact in all directions, *as many as are constantly nourished and therefore continue to live (hosa aei trephetai te kai zêi dia telous)*, so long as they are capable of absorbing nourishment.

It is to be noted that, as with the definition of life in II 1, we have no causal account of why it is that living things continuously nourish themselves – for that we have to wait until nutritive soul is defined in *de Anima* II 4, discussed in our next section, and then the discussion of connate heat, prepared in II 4, in *JSVM* (see below §4.4).

For the position presented here, there is in fact no need to deny the continuity of thought between II 1 and II 5; indeed the similarity of language would make it hard not to jump to the conclusion so many readers have jumped to. The peculiarity of my position lies in insisting on the nature of first actuality as nutrition; there seems no way one can duck the question *which* actuality is meant by first actuality. And it is clear that in II 5 nutrition and the growth it supports are on Aristotle's mind. For the alteration needed to move from something that has the (first) capacity to second capacity/first actuality is a change caused in generation by the progenitor (417b16-18); the first actuality of something growing, as an independent thing, is that of nutrition. Thus to grow into something with the capacity to perceive, nutrition must be taking place. This then is the first actuality, as suggested by our reading of II 1. As a bonus, we can offer a reason why there is neither in II 1 nor anywhere else a second actuality: because none such exists. In growing, the living thing does not acquire another, second actuality. Rather, there is only one actuality which continues; but in its continuation in the case of animals it gives rise to and preserves capacities for activities, inherent but undeveloped in the animal's actuality when it first appears.[84]

My interpretation is based on the idea of a natural life-cycle, discussed above in the Introduction and suggested in *de Anima* II 1 by the preliminary definition of life: self-nutrition, growth and wasting away. The general account is general because it captures things at a stage in their development before they are differentiated. Thus, although in this way the definition captures entities, as it must do as a definition, it captures them only vaguely; and it offers no explanation, which a conclusive definition will have to do.[85] Since the enquiry into living things is completely general, it remains to be seen how this generality can be preserved, if the differentiated living things as such are to be captured.

3.3 Growth and nutrition

In the last two sections we have reached the conclusion that in order to read *Parva Naturalia* we need to be aware of what has already been said in *de Anima*; and that the most general definition of soul is very closely connected to the nutritive soul. It is thus to the latter that we must now turn our attention. In doing this, we are merely following Aristotle: once the most general definition of soul has been given and discussed in II 1-3, he turns to the nutritive soul as the soul that is present in all the others.[86]

The account of nutrition given in *de Anima* II 4 is introduced as one of the relatives or 'objects' of the different faculties of the soul. These relatives are necessary to an account if we are to describe these faculties.[87] If you wish to understand perception, then you must understand perceptibles; if nutrition, then food. To start with, we must deal with the things relative to these activities, thus Aristotle starts with food.[88] His discussion is guided by the idea that when you are explaining a capacity, you should first look at its activity[89] partly perhaps on the grounds that activities or changes, unlike capacities, are accessible to inspection. This methodological reason is, however, grounded in what a capacity is, namely a capacity *for something*, such as a change or an activity. Since any capacity is capacity for something, the correlated activity or action provides a sure lead in tracing the capacity. And, since these capacities always relate to something else – its 'object' (*antikeimenon*), involved in the activity or change for which it is the capacity – this something else must be considered first. So, in the case in hand, Aristotle is led to consider nutrition (*trophê*), here taken in the sense of food: the capacity of animals to nourish themselves has as its activity, active nourishing, and is essentially related to food.

Several obstacles have to be overcome before actually tackling the question of what food is: first a distinction is made about what the nutritive soul does; secondly an account is given of the way the soul can explain what living things do. First, the distinction: Aristotle divides the function of the nutritive soul into two: to reproduce and to use food.[90] What is the connection between these two functions? Why do they both fall under one kind of soul? The simple answer is that they are both concerned with preservation. The individual must feed to survive; and it must reproduce if its kind is to survive.[91] Since the final end is in fact to produce another like individual, Aristotle even thinks that this soul should be called generative soul.[92] This can be taken simply to mean that if we arrange feeding and reproduction in an order according to which one can serve the other as the end, reproduction must take precedence as the final end. Nutrition of the individual is subordinate to the end of preserving the kind. There is, however, also a particularly Aristotelian reason for the connection between the two functions of nutrition and reproduction, namely in the matter involved. Reproduction depends on the matter that feeds the living body.[93]

And for the second preliminary: we have already seen how the ways of explaining things, Aristotle's causes, are related to his conception of the nature of something: nature serves as matter and form, but also as moving cause and the end achieved by this change. And since such natures especially belong to living things – they are most obviously independent and organised for an end – it comes as no surprise that the soul, which after all is the form of a living thing, also serves as cause or principle for the living body.[94] It is explanatory as the form, or substance, as it is expressed here – the cause of its being, i.e. its life, since being is living for living beings, but also as the end of the body: natural bodies are all tools of the soul – a formulation that reminds us of the most general definition of the soul in *de Anima* II 1.[95]

Both of these explanatory modes have clear if unexpressed relevance to the topic in hand – the soul is the cause of existence or life of living things, since the soul is, among other things, the actuality of nutrition, and the parts of the body serve this actuality. However, for the last kind of explanation as source of change, the soul is expressly related to growth, and not merely locomotion – perception also is included as 'a kind of alteration':

> Nothing perceives that does not participate in soul, and it is likewise with growth and wasting, for nothing wastes away or grows naturally, if it is not nourished, and nothing is nourished that does not share in soul.[96]

Given that the nature of something is a source of change and rest, and that soul is such a nature, it follows that the soul is a source of change and rest;[97] and since as we have seen the primary natural change in living things is growth, the causation of growth and the related function of reproduction is particularly closely connected to the nature of the thing: in our chapter, Aristotle calls these functions the most natural of all.[98] Thus this introductory passage on the soul as a cause is not an excursus, but germane to the treatment of nutritive soul, insofar as it is responsible for growth and decay as changes.[99] There are a number of striking things about this passage. First, the nutritive soul is being made responsible for the changes in the living thing, *including wasting away*.[100] That is to say, both form and matter are necessary for the explanation of this as of any other part of the cycle. Even if we say that soul fails in the latter half of life, we do not mean that it fails completely, nor that it would be better, given the living thing concerned, were it not to fail. (We can see here some of the dangers of teleology, in that we might be tempted to say that soul fails in its responsibility, or some such phrase; but it is not, in this instance, an agent with a choice.) One of the main tasks that face us is to show how it is possible and indeed necessary for wasting away to happen in a living body. It would appear from the passage cited above that both growth and wasting away are natural – parts of the natural life-cycle. And of course, while wasting away is happening, nutrition is still happening too: only living things waste away. The way to solve this problem is to look at the account of nutrition. This will show how form and matter are related, in that the matter is always present in some quantity, how great this quantity is determines which stage in the life-cycle is present. Once we understand that soul plays a part in explaining wasting away, we will understand how it is that only living things waste away.

The second thing this short passage indicates is that nutrition only occurs along with soul, as for perception: in this sense soul explains each of the activities. That soul is needed to explain growth is then argued in the following passage.[101] The first opponent is Empedocles, who is taken to task for thinking that the elements with their inherent movements up (fire) and down (earth) are respectively responsible for growth in these directions. The main reason that Aristotle argues against this view of growth is that living things would apparently be torn apart if there was nothing such as the soul to hold them together.[102] The second naturalist to have erred, also in seeing no controlling factor for growth, although anonymous is in all probability Heraclitus – we have already noted his influence on Aristotle's thought on living things, and will find more as we proceed. Heraclitus is said to make fire the cause of growth since it is the only one of the simple bodies to feed and grow. But as Aristotle points out, a fire will grow as big as it can, given the fuel available, and so, although he concedes fire's co-responsibility for growth, the main responsibility is the soul's:[103] here we see the idea that there is a boundary inherent in living things, and this is provided by the soul. There is a natural limit to growth, as is required by the idea of a life-cycle.[104] One might be tempted to take this limit purely quantitatively: few humans are more than six and a half feet tall. But clearly the quantitative limit is reached in a certain period. To put this back in context: both growth and wasting away are laid at the soul's door: the soul determines *when growth is complete and so when wasting away can begin.*

Many problems suggest themselves, foremost among which is the following: if the soul is immaterial, how can it have an influence on material things? In this case, Aristotle wants the soul to impose a boundary or definition on growth, or more generally on change. In this sense the soul is the cause of change, i.e. of the natural, characteristic – indeed necessary – changes of living things. Yet changes here are attributed to the elements fire and earth; how are we to move from the elements, with their boundless changes, to the bounded and living changes caused by soul? To start our answer to this question with the criticism of Empedocles: supposing fire rises and earth sinks, one way of reaching a compromise between the two would be to mix them. In this case, the elements would limit each other, and most importantly, the way they would do this is in their changes or motions: fire would heat earth and so make it lighter, itself sinking, but raising the earth. This would not be a case of the soul limiting the changes of elements; and it would be unclear how they would engage in this unnatural miscegenation. But suppose that there exists a certain mixture of these elements, so organised that it engages in the activity of nutrition and generation: then this activity, using the changes of the elements, would secure the perpetuation of this form of arrangement; and the way it would do so is through the changes being arranged in a certain way. It is this idea that I have already suggested as a way of understanding how changes serve ends in Aristotelian physics: changes can be arranged in a certain way, and so bear a certain form. The way this conception is filled in by Aristotle is the subject of the next chapter; the reader has now in his or her possession a promissory note to the tune of a material account of living bodies.

Now that the preliminary obstacles are out of the way – the connection between generation and nutrition has been posited, the soul's responsibility for nutrition exposed – let us follow Aristotle by turning to nutrition.[105] The term (*trophê*, 'nutrition') in both Greek and English is ambiguous between the matter needed (food) and the process it is involved in (feeding).[106] If in the soul we have an explanation for the formal side of nutrition, it is reasonable that the material side will be covered by food. As suggested at the start of the chapter, Aristotle approaches the activity of feeding through its object, food. Yet, as will become clear, food is not something to be approached by leaving out the activity of feeding. For food is relative to what feeds on it. But the question is: how are they related? Obviously they are related in the process of feeding: but just what is the relation implied by this process? Two relations are suggested: they are opposite to one another, or they are the same, i.e. instead of being at opposite ends of a range of qualities, they are both at the same end. Thus the question to be answered here is whether contrary feeds contrary or like feeds like – does fire feed on fire or on water, assuming that where fire is hot and dry, water is cold and wet? The question is disputed because the like cannot undergo change under the influence of the like; all change is to contraries or middles; and the food is changed by what feeds on it.[107]

This is a specific instance of the general account of producing a change in something else and undergoing a change under the influence of something else in *de Generatione et Corruptione* I 7.[108] For the producer of the change must assimilate to itself what undergoes the change: the fire assimilates the kettle to itself by heating it. This requires that at the outset the two things possess two capacities or predicates from the same range: this warm, this cold. This range is the genus ('a certain amount of heat') shared by both things, so that they can interact; but of course if they were also at exactly the same place in the range, not just generically the same, but also specifically, they could not interact either: if the kettle is as hot as the fire, the fire will not heat it.

The solution to the question about nutrition is that we must distinguish between concocted and unconcocted food; when we make this distinction we can say that the like is fed by like, and by its contrary.[109] There is a difference between what has not yet been acted on by the nutritive soul and that which has. Before it is digested, it is contrary to that to which it will be added; but once it has been digested, it is like that to which it will be added. In other words, a change has taken place. Thus for Aristotle nutrition is, in the strict sense of the term, a process of assimilation: it makes food *like* the living body it feeds the food is changed and concocted into the living thing. Capacities are relative, a capacity to act relative to a capacity to undergo change, and conversely; and nutrition is relative to the body with a soul.[110] So food is relative, and without bodies with souls there would be no nutrition. It is easy to understand that food is relative to those living beings that feed on it, if a living thing must be able to make its food like itself. What a living thing can assimilate depends on that living thing. And, as a final remark, if food is introduced as the material cause of nutrition, we can see that, once it is assimilated to the fed body, this too can be seen as the matter in nutrition, as it were, the finished, formed matter.

Already in this discussion Aristotle has made clear that nutrition is closely connected to growth, a not unreasonable connection to make.[111] Provisionally, we can say that growth involves a change in quantity, whereas nutrition preserves the substance (*ousia*).[112] Through nutrition the living being continues to exist as what it is; growth is merely a change (increase) in quantity. Once this distinction has been made, Aristotle gives a definition of nutritive soul:

> Such a principle of the soul is the kind of capacity that preserves that which possesses it as such, and nutrition prepares it to act. Hence that which is deprived of nutrition is not capable of being.[113]

After the definition of nutritive soul he makes a threefold distinction of the factors involved in the process of feeding. This passage (416b20-9) is of great importance since it establishes the relation between food, nutritive soul and the body that is fed. That is to say, we learn more about the relation between body and soul from it. Lines 20-9 may be translated as follows:

> Since there are three factors, what is nourished, that by which it is nourished, and what does the nourishing, the last is the primary soul, what is

nourished is the body possessing the soul, and that by which it is nourished is nourishment. Since it is just to call everything in view of its end, and the end of this soul is to produce another [living being] like itself, the primary soul would be [on this reasoning] 'the soul productive of another like itself'. That by which [the soul] feeds [the body] is double, just like that with which the soul steers: hand and rudder – one thing changes and is changed, the other only is changed.

The three factors are:

(1) that by which something is fed is nutrition (food);
(2) that which is fed is the body;
(3) that which does the feeding is the nutritive soul.

In the account so far we have met all three factors; now we need to determine their relations to one another more closely. In the second part of the passage (25-7), a distinction about the kinds of instruments involved is added, which is made difficult by the state of the text.[114] From the first part, we can extract something approaching a grammar of nutrition. 'Feed' is being taken as a transitive verb, of which the subject is the soul, the object is the body, and nutrition (food) is somehow 'that by which' the feeding takes place: the soul feeds the body with food.[115] To do so it uses heat and cold in the body as the prime instruments by which the assimilation already described takes place.

So we have matter and instruments: a carpenter uses tools; but he also uses wood. Food is not the instrument of nutrition; it is the matter that the tools can work on.[116] So it comes as no surprise that we now have a description of food: it is that which can be digested.[117] Corresponding to this (passive) capacity of food, there must be the active capacity of the soul to digest: capacities for change come in pairs: parallel to a capacity to heat, there is a capacity to be heated.[118] This particular change, concoction, is caused by heat.[119] This argument forms a very clear example of the argument from the activity to the body, that is, to the capacity: all living things nourish themselves; nourishment requires heat. So all living things have heat. Now the tool, heat, is mentioned. Of course this should not be taken to be the end of the account of the tool of nutritive soul: living bodies are organised in an elaborate manner so that they can nourish themselves.

And it is to some such further treatment of nutrition that II 4 refers us. This is one of the references which have often been taken to mean that we are missing a work by Aristotle called *On Nutrition* (*peri trophês*).[120] The question whether there was a separate treatise on nutrition is made more difficult to answer by the fact that, in the nature of things, the rest of natural science has a lot to say on the subject. One important source for our knowledge of Aristotle's views on nutrition is *Historia Animalium* VIII

1-11 on the different modes of feeding (*trophê*). Nutrition (*trophê*) plays an important role also in the *de Generatione Animalium* because of the primacy of growth in the process of generation and of the nutritive soul. It is also relevant that Aristotle sees semen as highly concocted food and that the foetus uses the mother as a source of food. In the *de Partibus Animalium* nutrition appears as the function of the heat in the heart.[121] We will meet it again in this role in *JSVM*. The subject of nutrition enters into several treatises of the *Parva Naturalia*; this allows us to entertain the possibility that the reference in *de Anima* II 4 is in fact to the *Parva Naturalia* as we have it. Nutrition plays important roles in sensation, since there is an organ of perception for food, and also in the explanation of sleep and waking. Nutrition is particularly active in sleep.[122] *JSVM* also deals with the topic of self-preservation, insofar as we are concerned with the change over time and preservation of the natural heat necessary for nutrition. Despite the inconclusive nature of the evidence for a separate treatise on nutrition, it remains possible that it was written.

Growth was grouped with decay and the prime of life as characteristic of living things. It was also contrasted with nutrition as being about quantity. To understand what is meant by these terms, and to understand more closely how they are connected to nutrition, let us now turn to the account of growth in *de Generatione et Corruptione* I 5. Aristotle says that three conditions must be fulfilled when something grows:

(1) Every part of the thing that grows must increase in size;
(2) Something must be added to it;
(3) The thing that grows must be preserved even while it is growing.[123]

The next step is to point out that things that grow are 'double'.[124] This is to be seen in the context of his idea that natural things are double. 'Flesh' is not just matter or just form: it is both. We then have the question whether it is the form or the matter that grows. The answer is that it is possible for something to grow with respect to form when something is added to it, but not with respect to matter. With respect to matter, things grow by the accession of something, and something flowing out (going beyond condition 2 above). He compares this to measuring water with a single measure: what comes to be is different. The measure measures different lots of water, so the water comes to have and then loses this measure, i.e. form. The form on the other hand is the same (condition 3 above).[125] The form increases in size by accession to each of its parts (condition 1 above). Although he illustrates this concept of growth using a hand, and uses the words 'figure' (*schêma*) and 'form', we should not take only the shape of the parts to be meant. As he says, proportional growth is clearer in the case of an anhomoiomerous part such as the hand. But clearly growth has to apply to homoiomerous parts as well, for anhomoiomerous parts such as hands grow through growth in the homoiomerous parts[126] and since flesh as such has no shape, we have to understand its form in some other way.

This answers the question: what remains in growth? The form remains, and hence is what grows. It also answers the question how growth can be in every part of the growing thing. The whole form grows and the matter does not remain. What is it that is added in growth? What accedes is in capacity that which it accedes to.[127] Thus something that becomes flesh has the capacity to become flesh, while it is actually something else. This is the food that changes from being actually hay, for example, to being actually horse flesh. In the account in *de Anima* there is great emphasis on this change in the food. In *de Generatione et Corruptione* it gives rise to the only mention of the soul in the account of growth.[128] Obviously, it is the soul that is the centre of the account in *de Anima*, namely as that which causes this change in the food. The form can be taken as what defines the thing that grows – the flesh, bone or whatever.

As we have seen in our discussion of *de Anima* II 4, Aristotle distinguishes between growth and nutrition using a distinction between what something is and how much there is of it.[129] Insofar as that which is added is a certain quantity of (e.g.) flesh, what happens is growth. Growth is, it will be remembered, in respect of size.[130] But insofar as the food is only potentially flesh, what happens is nutrition. This is what it means to say that nutrition preserves the substance (*ousia*). Flesh remains flesh because of nutrition. This process takes place throughout life, irrespective of whether the living thing is growing (increasing in size) or decaying (diminishing). As a quantity, some flesh *grows*; but *qua* flesh it is *nourished*.[131] Hence it is nourished as long as it is preserved, even when it is decaying, but it does not grow always. He continues:

> And the [process of] nutrition is in one way the same as growth, but in its being it is different. For insofar as that which is added is potentially an amount of flesh, it is thus far productive of growth of flesh; but insofar as it is only potentially flesh, it is nutrition.[132]

Growth and nutrition may be the one process, merely considered from different points of view. That both growth and nutrition are being considered in the chapter is clear from the comparison of the capacity to grow with a log fire.[133] When the logs are lit, we have a new fire, like a case of coming to be. But when one puts logs on an existing fire, that is like growth. Yet one could say that when you put logs on a fire you do not necessarily make it bigger: you could just be keeping it alight, i.e. feeding it. Feeding preserves the fire. (Note that in a fire, logs are consumed; hence the need for more logs.) Thus in these passages Aristotle has both growth and feeding in mind.

There is, however, one apparently decisive factor against this reading: nutrition is not part of coming to be whereas, obviously, growth is. Thus growth, and not nutrition, has a place in a work on coming to be. But this impression is misleading; for in nutrition food *comes to be* the living thing; and no new living thing comes to be without being nourished. In some way nutrition is more fundamental: when you are growing, you are being fed, but this continues throughout your life. Yet the point about growth, and indeed decay, is that nutrition is always a quantity.[134] Either too much or too little food is provided for the body. If there is too much, it grows, if too little, it shrinks. The stage in between, when there is a balance between provision and need, is the prime of life: it is only expressible in quantitative terms. Just enough food is provided by nutrition as is needed to preserve the body. Nutrition (food) is always, i.e. at any stage of the living thing, a quantity.[135] If you regard the growth of merely flesh, rather than this particular amount of flesh, then that is an abstraction. It will always be some quantity of flesh that is nourished.

So what has emerged from this discussion of nutrition and growth? Nutrition is a process that can best be understood by looking at its activity – this involves the soul as a moving cause assimilating nutrition to the body. The grammar of nutrition, as I have called it, is that the soul is the subject of 'nutrition', the body the object and the wherewithal is provided by food. But nutrition is not only *of* the body, it is *by* the body: the body not only benefits from nutrition, it also provides the tools to do so. Since these tools are body, they are preserved by the process to which they are instrumental. (Food, it is to be noted, is parallel to the carpenter's wood, not his saw.) The process of nutrition is always quantifiable, thus allowing three stages to be distinguished: where more is produced than the present body requires for its preservation, where the same amount is produced, and where too little is produced. So we are, at least schematically, in a position to understand the stages in life – growth, prime, decline.

In the case of animals, feeding provides matter, and so the capacity for change. In the case of plants, it provides only raw materials. They get their energy from outside. Aristotle does not seem to make this distinction between raw materials and energy. But the connection between them in his account can be made a little clearer, as follows. Let us take it that energy is what enables some thing to do some work. For Aristotle, such work would be something like digestion, motion and perception. These are things that living things *do*, in his account. So what does he think they need in order to do it? Primarily, they need heat.[136] There must be heat so that the living being can nourish itself. As we have seen in nutrition, new tissue is prepared out of food. That is to say, there is recognition that raw materials are needed and transmuted into the body of the animal. But

57

what about heat? It seems that Aristotle regards the heat of the body as like a fire that has to be fed: not only does it prepare the raw materials, but it also feeds on them and so preserves the warmth of the living body.

Body is an instrument of nutrition, as well as its beneficiary: this confirms a connection with the definition of soul as the first actuality of a natural, instrumental body. And since heat is prime among the instruments of nutrition, an important aspect of a treatment of living body will be the heat needed for nutrition. In turn, this heat must be somewhere in the living body, and so must be located. In the *de Anima*, Aristotle does not go into the details of the organs concerned.[137] They are of course *body*: any further account of nutritive soul will involve body. And the *Parva Naturalia* is where we look for an account especially involving both body and soul. But before moving on to these further tasks, we must approach an objection to Aristotle's view of soul. For the soul has emerged as the principle of change, especially of nutrition, as including the processes of changing food into living body; what is nourished is the body, which in turn effects the changes. That is, as it were, the positive side of the picture. We must now turn to the question of what decays. Perhaps there is no surprise in saying that the body decays. But the reader may be forgiven some irritation at being asked to believe that capacities of soul, such as perception or nutrition, do not decay. And that seems to be entailed by saying that the soul does not decay. In other words, what is the case for saying that soul changes body (transitively) without changing (intransitively)? On this turns the possibility of maintaining the distinction between soul and body. As an activity, nutrition would seem to contain a connection to matter, the food it works on, the tools it works with. The next section investigates the difference between body and soul, based on the idea that body is subject to change, where soul is constant.

3.4 Natural form and natural matter: what perishes and what does not

As we have seen in the last section, nutrition is a process by which living things continue to exist. This fits into the way in which we have taken Aristotle's teleology, namely as a kind of order supported and achieved by means of change. Nutrition is itself not a change of the living thing, but involves change of the food – assimilation to the living thing. Yet there are problems here; for if the living thing is preserved by nutrition, the question arises of how it is possible for it *also* to go through a life-cycle, including its wasting away. Part of the answer to this question has been given in the last section: nutrition always involves some *amount* of food, so, since this quantity can vary, nutrition can take place in a greater or less quantity, so explaining the different stages in life. This explanation assumes that nutrition of the living thing is something arrived at by change, and so itself not subject to change. Just as in the process of building a house, the builder, as such, does not change, but rather the matter in which the house comes to be, so it is the food that changes – being that which is changed into, assimilated to the living thing – and not the living thing. Applied to the process of decay, the idea is that the activity or form does not change, but rather the matter associated with it.

Aristotle argues for the thesis that forms do not come to be in the *Metaphysics*.[138] The argument there is general and not based on the peculiarities of living things: if every case of coming to be ends in the presence of a form in matter, then the supposition that the form and the matter come to be will result in an infinite regress. The basic supposition is that, in order to analyse change, one needs a form and some matter which can have or not have this form. Instead of the form or the matter coming to be, it is rather the form in the matter that comes to be. For if the form comes to be then it, too, would have to have components, i.e. matter and form, and these too would have components, and there would be no reason to stop this process of analysis. Why, one might ask, is Aristotle so against the idea that the factors in change might themselves be infinitely divisible?

He gives no indication here, beyond making it clear that forms or matter cannot be further divisible into form and matter.[139] Given the view of soul suggested in this chapter, we are in a position to support this idea in two directions. On the one hand, if living form were divisible – meaning divisible and able to retain its identity – we would have problems with the idea that soul is an actuality. For an actuality is a unity, comparable to a point, in that it is either there or it is not. It is not something that comes about by addition, until it is complete.[140] It is complete from the start. On the other hand, it can be made at least plausible that a process such as nutrition must be complete to exist at all. After all, suppose that at any time, the functioning of some certain group of organs is necessary in some certain kind of living thing for nutrition to take place. Then, trivially, without the functioning of one of these organs there will be no process of nutrition – not merely partial nutrition, but no nutrition at all.[141]

Given that living things are composed of body and soul, when we look for the cause of long life we can ask which of these two is involved. If we understand what happens in ageing to be a kind of change more specifically, a kind of perishing – the question is what it is about living things that is suited to perishing. Aristotle asks this question in the second chapter of *LBV*. Before turning to that account it will be useful to see how that work is introduced.[142] As usual, Aristotle begins from a discussion of the puzzles involved:[143] the main puzzle is whether the same or different causes are responsible for animals and plants being long-lived and short-lived.[144]

In *LBV* 1, Aristotle first turns to animals. In the first instance, he asks which entities have different lengths of life: different kinds and different individuals within the species.[145] The examples of kinds (*genê*), 'horse' and 'man', and species (*eidos*), 'man', suggest that *genos* and *eidos* are being used interchangeably here.[146] So we have two simple contrasts – between members of a species, and between different species. It is in an important sense a comparative study. We are dealing, as one expects in the *Parva Naturalia* (see above §3.1), with phenomena that are shared either by all living things, or by those in one kind (*genos*) or one species (*eidos*). It is about the cause of longevity not only in animals but also in plants. On the one hand, horses and men differ in the length of their lives; but both have specific life-spans. On the other hand, Aristotle thinks that those 'kinds' of men living in hot places are longer-lived; and those people living in the same place live different lengths of time.[147] The topic of locality, where something is and hence its surroundings, is one that will prove important later.[148] For now, it is enough to note that finite substances must be considered in relation to their surroundings, if we are to understand their behaviour, even if in the last analysis this behaviour is natural to them.

The second chapter of *LBV* moves onto another level: rather than viewing the different kinds of living things that exist, and asking which are long-lived,[149] he asks about the constitution of all living things: he says we must grasp what it is in natural composites that is easily perishable and what is not. But he does not go straight on to discuss this question. In fact, he does not even begin to do this until chapter 4. Instead, *LBV* 2 tackles the question of the perishing of living beings in terms of whether their soul has its own way of perishing, besides that of the body it is in. As one would expect from the arguments in the *Metaphysics* and elsewhere already cited in this section, he denies that it does. It follows that since such composites do perish, their perishing must be due to their body. But how are we to understand 'body' here? Close attention to the text is needed.

Although we have already met food and the tools working on it (see above §3.3), this chapter is our first real contact with Aristotelian matter; and it is not by chance that this contact takes place in a chapter on soul. For body is approached through the soul, as relevant to the functions of living things. Nonetheless, matter has here a distinctly non-living appearance, and as we will see in the next chapter, matter does indeed have a non-living incarnation: as the elements fire, air, water, earth.[150] Aristotle says it is reasonable for natural composites to share the natures of the elements of which they are composed.[151] The way elements are introduced here allows us to take a relatively non-technical line: they destroy one another. This can be taken to mean simply that, for example, when the hot in fire works on the cold in water, the water perishes. The second step is that this perishing is not restricted to the elements as they are when unmixed, but also continues in bodies constituted of these elements: 'things akin to them participate in their nature.' At the least this must mean that things constituted of elements are characterised by the same capacities as the constituent elements (their 'nature'). These capacities here are contrary to one another.[152] Put simply, things made of earth, water, air and fire share the capacities of these things. So that, assuming living things are constituted from the elements, living things will share the nature of elements, and thus perish because of the contrary capacities.

Two thoughts might occur to the reader at this point. The first is that the elements, understood as being opposed to one another in possessing contrary capacities, do not so much explain the eventual perishing of living things, since the contrary capacities would work against one another, but rather make clear why this conception of living body is untenable; for anything so constituted would tear itself apart. This is a point we have already met in the last section[153] – Empedocles made the motion of the elements responsible for the growth of things, Heraclitus made fire responsible, and Aristotle wanted soul – in the one case to prevent the body being torn apart and in the other to limit growth. What this piece of argument makes clear is the need for a factor to hold things together, or limit their change. Soul is meant to act as a stabilising factor. In some way, nutrition, an activity, neutralises the effect of the elements – so the heavy does not sink, or the light rise, or fire burn untrammelled.

Nonetheless, fire serves as co-cause of nutrition. This will serve as an introduction to the second point: are the elements not being made responsible for decay in *LBV* 2? Is it not clear that living things derive their qualities, and hence their destructibility, from the elements? The nature of the 'fire, water and things akin to them' (465a14-15) would seem to be at least their contrary capacities, in consequence of which they destroy one another; they are then responsible for the presence of this capacity in the living things in which they are present.

My answer to this has two prongs. The first, and more straightforward, goes as follows. Aristotle's claim that 'beings share the nature of the simple bodies, each of the others being out of and consisting of these simple bodies'[154] means that sharing the nature of the simple bodies means possession of their capacity to destroy one another. For example, just as fire might destroy water, an animal can heat things and hence destroy (replace) cold because it is a hot body: contact with a warm animal can melt ice. This need not mean that it is fire actually in the animal that does the melting. We must note that while fire and water are 'elements', the things akin to them might well be hot and wet stuff of which living things are composed. Being akin to an element does not mean being an element, but having one or both of the capacities also present in the element. Being akin to fire means being hot and moving upwards, if nothing is in the way.[155] As we shall see, both these aspects of hot stuff play a central role in Aristotle's account of breathing.

The second prong consists in asking why fire and the other elements are in living things. This 'why' has both a functional spin: what function do they serve? as well as a connection to nutrition: how do they come to be part of living things? As we have seen, at least fire serves the function of nutrition in living things; and it is clear that the way in which elements come to be in things is through nutrition assimilating them to living things.[156] The effect of this prong is to insist on the presence of soul in any explanation of living behaviour: only when we have understood how some part of the living system fits into that system will we have understood how the part operates in the system.

LBV 2 asks about those parts of living things which are easily perishable *and which are not*. This fits perfectly with our insistence that *life and death* are to be treated together: *LBV* is a treatise on the length and shortness of life, both on what prolongs life and on what shortens it. While Gill does not deny that parts of living things such as flesh are to be explained in Aristotle's view by the functions they perform, she insists that ultimate matter explains perishing in mixtures.[157] But clearly this is not the whole story. The point is that you cannot isolate matter as a factor in living things; it always has to be seen in relation to the function that it fulfils. Matter is not the villain of the story, only causing perishing; it also causes life: there is no (mortal) life without matter. On the one hand, an organ's qualities or capacities enable it to function – the heat and elasticity of the lung, for example. On the other hand, it is these capacities that leave it susceptible to change and hence perishing. Thus function must form part of the explanation of the perishing of living things. This perishing is related to, but not reducible to, the perishing of elements into one another. The same capacities – hot, cold, dry, wet – are involved in each case; but, in living things, these capacities do not belong to elements but to the living things.

Another problem is how bodies made of different elements can form a single unity when the elements are characterised by simple qualities in pairs, or by the directions they tend to move in.[158] This is relevant to mortality because if the elements have capacities to move to their proper places, and these capacities are in mixed bodies and *not* in their proper places, then we would seem to have found another extraneous reason for the necessary perishing of living things – extraneous, that is, to the kinds of thing they are. In fact, this line of questioning is a repetition of the problems with the reductionist reading of *LBV* 2. A two-stage answer to this argument can be given. First, the immediate, and hence best, explanation for the behaviour of living things must involve the matter they are made of immediately, rather than some matter that may be involved at several removes. For 'ultimate' matter might explain anything (in some sense) – water goes into the composition of many things. But a complicated fluid such as blood, for example, explains the relevant functions of the body much more nearly.[159] Secondly, we must then explain how mixtures such as marrow and bone come to be out of elements. The problem is simply that the capacities in mixtures no longer belong to elements, but to the mixture. So, even if the capacities cause the perishing of the mixtures, the elements do not.

LBV 2 is permeated by a contrast between things constituted by nature and those which exist by synthesis from many things. There are two examples of the latter: houses (465a18-19) and knowledge (along with ignorance, sickness, health: 20-1, 22-3). With a house, synthesis can be taken in a technical sense developed elsewhere by Aristotle.[160] In this sense two things can be put together, if they do not then cease to exist as actual things. For example, bricks, when made into a house, do not cease to exist as actual things. That is why such a thing as a house is said to consist of many things.[161] As a contrast to this we can remember the assimilation of food described in the last section: when assimilated to a horse, hay ceases to exist as actual hay.

The second case of synthesis is simply the putting together of an accident with a substrate that can survive without it: knowledge has its own destruction, e.g. forgetting, so the soul in which the knowledge existed can continue to exist when the knowledge has gone. Even if house and health seem to be very different kinds of things, a similarity can be found. The way that the things come to be together is decisive for the kind of unity they form: naturally, i.e. by growth, or by the exercise of some art. This provides the clue to why house and knowledge are taken together: both contrast with the things constituted by nature.[162] So houses and knowledge can either perish their own particular perishing, or else they can perish through the perishing of the thing they are in, their 'recipient'.[163] Knowledge can be forgotten; or else the possessor of the knowledge can die. Either way, the knowledge is then no more.[164]

From this argument, he continues, one can draw a conclusion about the soul: if the soul were not in the body naturally, but like knowledge in the soul, there would be a separate destruction of the soul besides that which happens when the body perishes.[165] The soul neither has its own peculiar perishing nor, and consequent on the first point, does it have a substrate ('recipient') that outlasts it, since when the living thing dies, its body and its parts are no longer something with a capacity to live.[166] Instead, the soul perishes when the body perishes. The argument[167] seems to have the following structure. Things can perish only for one of the following two reasons: they have their own peculiar perishing, or their substrate perishes. The first does not hold for soul, and is excluded as being obviously false. So only the second possibility remains, with which the chapter had anyway started. Soul perishes when the living body it is in perishes through the action of contraries on one another.

The chapter does not directly tackle the main question of the treatise but prepares the way by showing that there is no independent perishing of soul – independent, that is, from the perishing of the capacities of the body in which the soul exists. The thrust of *LBV* 2 is not to ask: do elements destroy living things? but rather: does soul have its own perishing? This possibility must be excluded, so that Aristotle can turn to the relevant qualities of the living being. *LBV* 2 does not restrict itself to inanimate mixtures, because all things with the capacities of elements perish. No reduction is involved, since the proximate matter of living things, *body*, is hot and wet. The idea that vital heat alone is responsible for length of life gains plausibility from the fact that function is not discussed in *LBV*.[168] But if one can show that such a discussion of function is part of the background, then Aristotle's theory of living things can be seen as a unity; the idea that matter alone causes or explains decay suffers from not seeing Aristotle's enquiry in context, and I have been at pains to show how this context fits together. *Parva Naturalia* and *de Anima* investigate both matter and form; which functions are fulfilled in living beings; and how they are fulfilled. Living a long time is not, of course, a function like perception or locomotion; but *LBV* 2 is the first approach to the question: what is it about living systems that causes their ageing? The answer is body, not soul. But it is still ensouled body.[169] The enquiry in the chapter is about body and soul together, and asks which of them is responsible for decay. Aristotle could not even ask this question if his science of life were divided into 'psychology' and 'physiology'.

LBV 2 argues that because the soul is naturally in the body, it is the body, as connate with elements that destroy one another, that is responsible for the perishing, and so longevity of living things. Before we go on to consider in more detail which living things are long-lived and why, the final section of this chapter looks at the location of the soul in the body. This will go some way to filling the gap Aristotle explicitly leaves on the subject of the partnership between body and soul.[170] In the course of this chapter we have seen the soul that is capable of producing its own body, whether by preserving the body it is in, or by producing a new individual in reproduction. In either case it contrasts with knowledge or house building. Let us now locate it in the body.

3.5 The seat of the soul

In this chapter we have pursued the soul from its most general definition, as the primary actuality of living things, and the connection between this definition and the nutritive soul, to the conclusion that, when something living decays or, more generally, changes, it is the body that will do this, not the soul. But so far the body has been present in our account merely as the elements, considered only in terms of capacities – hot, cold, dry, wet – which interact and either destroy one another or give rise to mixtures. That is a very abstract and general consideration of body; and, as we have seen,[171] Aristotle considers it a major mistake of his predecessors not to have thought about the kind of body that soul is in: what is it about the living body that gives it a form? And so we must now turn to the consideration of the living body, insofar as this is the locus of activity or actuality. Two ideas are combined in the treatment – first that the soul is an activity or an actuality which parts of the body serve, and secondly that the way parts of the body do this is by changes arising from their constitution and systematic interrelation. This shows how form is both heuristic – the functions of the soul guide the causal analysis of the body – as well as real, in that such souls exist.[172] The soul is in the body, insofar as the body possesses the capacities to perform functions that fit into the actuality as a whole. This is a first move towards explaining what it means to say that the soul is first actuality of a body equipped with organs.

In the opening chapters of *JSVM*, Aristotle locates the seat of the soul. After noting the connection with *de Anima*,[173] two requirements open the treatment (467b14-16):

(1) The soul is not body, nonetheless it is in a part of the body.
(2) This part must have power (*dynamis*) over the rest of the body.

Both these requirements fit closely with everything that we have learned so far about the soul – as the activity of a natural body with organs, soul must be what is served by, and so controls the parts of, that body; that is the point of looking for the soul in the body.[174] The second requirement is concerned with the idea of the body's organisation. Given that the body is involved in activities and has different parts, the question arises of how these parts are related to one another and so to the activity in which they are involved.[175] The search for a part which exercises power over other parts suggests that control is being thought of here as originating change in other parts: it is the power or capacity for change that the part possesses.[176] That this part has the capacity to bring about change in the other parts of the body entails that these parts are connected to the seat of the soul, and that they are of such a constitution as to be changed by this part.[177] That some such part has control over the rest of the body follows from the fact that activity is possible only through concerted changes in the different parts of the body, and, apparently, without control the relevant ordering of these changes will not happen.

The further argumentation, however, for there being *one* part in virtue of which the animal is both animal and alive has not yet been given and is very obscure indeed.[178] It gives one reason to think that Aristotle is here working with other, unmentioned presuppositions. The reasoning for the unity of animals' central organ goes as follows. Aristotle notes that one can say two things about animals:

(1) they are alive;
(2) they are animals.[179]

These two assertions are not equivalent since there are things covered by (1) that are not covered by (2), namely plants: plants are alive but have no perception, and it is through perception that plants are distinguished from animals. Nonetheless, he wants to assert that the part in virtue of which (1) is true of animals must be the same part as that in virtue of which (2) is true of them. The reason he gives is that it is impossible for an animal as such not to live.

Therefore, while that in virtue of which an animal is alive is one in number, it is several and different in its 'being'.[180] These ways of being are clearly to be identified with being an organ or principle of nutrition and being an organ of perception – that is, with two functions in contrast to the single, identifiable locality where they are situated. That in virtue of which an animal is alive is the principle responsible not only for perception, but also for nutrition, since everything alive nourishes itself, and an animal *as such* is alive.

But this is no argument for the unity of the parts in which perception and nutrition primarily take place; without some question-begging premise about the way in which nutrition and perception as actualities are connected, it merely amounts to a blank assertion that the principal organ of nutrition and sensation must be the same.[181] All that the argument establishes is that the part or parts in virtue of which animals are alive must perform two functions: nutrition and perception. Even if there are two functions, they can be performed in different parts; unless we assume that perception and nutrition are so connected that they must be performed by the one part. Nothing in the description of the soul has prepared us for the unity of the primary organs of perception and nutrition. There is no reason for a unitary central organ in the function of these parts of the soul. For example, there might be one primary organ for each kind of soul.[182]

Since no real reason is given for the unity of an animal's central organ, we can ask why Aristotle thought it unitary. If it is right to say that the being in view of which the central organ has distinct aspects is the activities, nutrition and perception, and the number in view of which it is single is that of its matter, the reason that the two functions are seated in the same organ has to do, not with the formal connection between nutrition and perception, but in their material connection. It is this connection that an investigation of the way the soul is in the body is best equipped to tackle. It can thus fill out the well known, but abstract formulations of *de Anima*,[183] namely that the kinds of soul are nested into one another: the basic soul, present in all mortals, is the nutritive soul, and this can form the basis for the perceptive soul, and this in turn for the intellectual soul.

This picture is compared to the way in which one can move from the simplest geometrical figures to more complicated ones. The simpler figures remain in capacity in the more complicated ones. A triangle is in a quadrangle in that it can be separated out from a quadrangle; in the same way the nutritive soul is in a perceiving being, and can be separated from perception. None of this tells us how the parts of the soul look in the flesh, as it were. Since this is the novelty *JSVM* introduces us to, the novelty of the unity of the organ of the soul need not worry us: it is based on material considerations, as yet not introduced into the discussion of the relations between the parts of the soul. The only problem this interpretation leaves us with is that it seems to amount to the fact that there is a material connection between the vital functions, and so they are situated in the same organ. The reader's reaction to this may well be divided. On the one hand, it is a relief to think that there should be a material connection between nutrition and perception, since otherwise it would be inexplicable how parts of the body, products of nutrition, are implicated in perception. On the other hand, it seems to require that Aristotle overlooked the possibility that two organs responsible for the different actualities could be linked, so both can perform their functions. Nutrition could be supplied to the central perceptive principle from another part (in modern life, technical near-analogues abound, a power station being an obvious one).

However, the possibility remains that in Aristotle's eyes there is a material link between nutrition and perception such that they both must take place primarily in the same locality. The relation between perceptive and nutritive soul is of great systematic importance, not only for the life of living things, but also for their death. If living things as such perform a variety of functions, one can ask if they have to perform all of them to be alive. Michael sees that a connection between the two kinds of soul is necessary since there is no animal life without touch, and if perceptive soul were separable then animals would live without it, and just have nutritive soul.[184] This argument does not, however, prove that both souls must be seated primarily in the same part of the body, merely that every animal has both: the argument asserts that no animal can be alive without perception. As far as this argument goes, the central organ, and hence that of touch, could be, for example, in the head, on condition that the animal has to have it as long as it is alive. In fact, we speak of an animal without sensation, in an *almost* perfect Aristotelian manner, as a 'vegetable'. Nonetheless, in the natural state, both the centre of perception and that of nutrition must function for the full life of the being concerned. Together they do form a unitary system, without being located in one place.

A further point concerns the specialisation of organs. One reason for the unity of perceptive and nutritive principle might be that it is simpler this way: nature makes things in the best way possible.[185] This principle can be detected here, even where we are dealing with an argument from the phenomena, in the connection between the unitary organ and higher development. As we shall see, in Aristotle's view there is a close connection between the presence of a unitary central organ and the higher organisation of sanguineous animals. This idea can perhaps be approached through the way he thought souls were localised. The question he seems to be asking is: is the principle of life in one part or in all parts? There is a spectrum of possible answers, ranging from 'everywhere in the body' to 'in one part of the body'.[186] In the first case, any part of the body might live, if separated from the rest (some plants can regenerate lost parts, in Aristotle's view). The latter case is that of animals: they are centralised, because there is one principal seat for their soul.

There may be a last reason, connected to the previous one, why he is so set on having a unitary organ for both perception and nutrition: he has his eye on the heart throughout this supposedly general account of the seat of the soul.[187] In other words, he is guided by what he thought of the heart even when discussing all living things, including those without a heart. For, as the source of the heat necessary to all vital function, it might have seemed a good central organ for perception as well as nutrition. He also thinks that there is no need for soul in each part, only in some source, i.e. principle.[188] That is to say, greater organisation and specialisation of the living body are possible, so that the vital function is concentrated in one place in better organised living beings. Instead of having the nutritive principle throughout the body, sanguineous animals have it primarily in one place; and this is a sign of greater organisation. In the *de Partibus Animalium* Aristotle makes clear that he thinks one source of control, where possible, is best as a general physiological principle, on the level of organs as well as of the whole living being.[189] The best organisation is, in his eyes, that of the sanguineous animals. The justification of this principle is another matter; it is certainly at play in this discussion.

So much for the argument, or lack of it, that there is one central organ in animals. Its influence is anyway decisive on what follows: the search is on for a *single* central organ. This idea has force even in those cases where there is no organ, but merely a part, as in plants. Much of the attention is devoted to sanguineous animals, so we must be careful not to think that we are only dealing with nutritive soul: both nutrition and perception must continue if animals are to live, in Aristotle's book. It is, furthermore, because they are necessary that these two functions provide the first two criteria from functional structure for localising the central organ. The investigation of living bodies takes place with the activities of the soul as guiding principles. Living bodies have three dimensions, and Aristotle thinks that each is significant in the organisation of the most developed living beings. It has been observed by scholars that Aristotle attaches great importance to the middle part of things, i.e. the part between the extremities.[190] The division into top-bottom, front-back and left-right plays an important role in his discussion of the body and its functions. He understands back and front in terms of the direction perception takes: forward is the way our perception is aimed, behind is the opposite, the middle lies in between.[191] He thinks that since the central organ of sense is where all actual sensations meet, it will be between back and front.[192] Thus the first bearing on the seat of the soul is carried out using the function of perception.

The second localisation of the central organ places it between 'up' and 'down', i.e. the top and bottom, namely as a seat of the nutritive soul (here, the nutritive principle). This means that it is the source of nutrition for the rest of the body. The top ('up') is where food enters. The ordering is relative to the body more precisely, to a vital function of the body: nutrition.[193] Although Aristotle's cosmos is ordered into 'up' and 'down', that is not what we are dealing with here. This division of the parts of the living being is not made 'in view of the whole surrounding' (*periechon*), but only relative to the places where food enters and waste leaves.[194] As often in the *Parva Naturalia*, he is very concerned to maintain the generality of his account: he thinks that plants are in this respect the reverse of animals: they take in nourishment through the roots, i.e. the lower part whereas animals do it though the mouth.[195] Thus they are the other way up to animals.[196] Teleology is guiding his enquiry at a variety of levels here (cf. above §2.3); first, in terms of the way things are well arranged, and secondly, in terms of how the body is organised so as to fulfil functions i.e. a good. The third way is that living things exist in a hierarchy of how well organised they are, and thus which functions they fulfil.

We are here dealing with complete specimens: they have all they should have, being what they are.[197] There is, of course, a particularly intimate connection between survival and completeness. If an animal does not have its vital organs, then it will not survive long. This fact was also used by Aristotle to locate the central organ in a second set of arguments. Chapter 2, using food and waste, pursues the identification of the central organ further by identifying the central region (in between up and down, back and front) in the largest animals as the chest.[198] For the purposes of Aristotle's enquiry here, there are three parts of animals: top, bottom and middle. As well as these, some animals have locomotive parts. Thus we have here a very abstract functional account of the parts of animals. He takes his indications from primitive experiments in which we are to think of animals as divided in various ways, and we ask which parts of them go on living.[199]

The generality of the account is preserved by the simple use of analogy: in other animals the central organ is in that part which is analogous to the chest. He finds confirmation for his view in the fact that insects can live when divided: it is the portion including the middle that lives.[200] More obscurely, he says that some non-insects too can live when divided, because the nutritive soul is several in capacity, though actually one. This applies also to perceptive soul, since such living beings have perception.[201] That is to say, the divided pieces will also display some form of sensitive behaviour. Such animals are like many animals grown together, unlike the more developed animals.[202] This points to their greater vulnerability: they have to preserve their integrity if they are to survive because of the systematic relation of their parts to one another.[203] Since it is the part that includes the middle which survives in a division of some animals, the principle of life must be in the middle. The same moral is to be derived from those cases in which division does not end in immediate death. Animals, in this case insects, unlike plants, can only survive division for a short while since they do not have the organs for preservation. That is to say, the divided parts cannot develop the necessary organs to nourish themselves.[204] The point, then, about parts, is their necessity to vital activity; the degree of complication in the arrangements to achieve this end has an effect on how easy it is to disrupt.

The whole of the rest of chapter 2 uses the example of animals that can live divided to show that nutritive souls and sensitive souls are one, but can be divided in these cases: they are many in capacity, i.e. they have the capacity to become many. The proof that nutritive soul is in the middle works using simple experiments – and these arguments are said to proceed 'according to perception', in contrast with the arguments 'according to principle (*logos*)' in chapter 4. In many cases, if the top or bottom of an animal is cut off, that part lives which includes the middle, i.e. the part with heart or its analogue.[205] This experiment enables Aristotle to conclude that the middle is necessary for any performance of the vital functions. If a vital function is exclusively or mainly carried out by one part, then that part is necessary to the survival of the being. If there are various parts fulfilling the vital functions, then *all* such parts are not necessary to the being's survival. Unlike such living beings, the higher animals have 'one nature and not several':[206] they are one and not composed of many. That is to say, they have no capacity to become several living beings. The peripheral parts of higher animals have no nutritive soul, but an 'affect' (*pathos*) of it: their parts are not independent, and so of course have no actuality of their own, but contribute to that of the whole body. This means that the parts are under the soul's control in that it must prepare food for them, which they can then finally use. They have an affect of the soul in that they are affected or changed by it: the soul nourishes them. That is to say, while all the parts are alive, they are alive by reason of their relation to the nutritive soul with its principle in the central organ. This is the systematic connection between this part and the rest of the body that both constitutes the greater complexity of the animal and causes its greater vulnerability.

In the case of things that are divided up, Aristotle is looking for that which makes them live, just as he is in the complete animals. The central point to be noted is the connection between the nutritive and perceptive soul. The divided animals do not live because they have no nutritive principle left, and food is necessary for preservation. Animals must have nutritive soul to survive, as well as the perceptive soul they must have to be animals.

Chapter 3 continues the arguments from perception from chapter 2, but now based on the way things come to be rather than on vivisection, dealing with plants (468b16-28) and sanguineous animals with nutritive soul (468b28-469a10) and then with perceptive soul (468a10-23). The supposed fact that plants grow from the middle provides reason for thinking that the middle is the part that controls or is the principle of growth.[207] It might seem strange to think that plants grow from the middle, when it appears that it is the shoots that grow, i.e. they grow from their extremities, and not the middle. However, if you consider the plant as a whole, you might think that the parts grow proportionally out from the middle. Even today we are familiar with the thought that a tree has roots extending below ground as far as the trunk and branches above ground. There is then a kind of symmetry about the middle of the tree. One might also think of the rings added to the trunk of a tree, also symmetrically from the middle. And symmetry along various axes is an important part of the study of plants. There is no suggestion in Aristotle that there is any organ at the middle of plants that is responsible for their growth in the way that the heart is meant to be in sanguineous animals.[208] Clearly, where no organ is present one can speak of the seat of the soul only in a weak sense.

In chapter 3 Aristotle derives grounds for the predominant place of the centre from the process of coming to be, and these provide the clinching argument in favour of the heart as seat of the soul.[209] The most impressive reasoning for a central location of a unitary central organ is anyway concerned explicitly with the heart. The reasons for thinking that the heart is the central organ of animals are:[210]

(1) The heart is formed first, and so can serve as the principle of growth of the rest of the animal;
(2) The heart is the 'source' of blood vessels carrying food to the rest of the body.

The conclusion is that the nutritive and perceptive soul of sanguineous animals must be in the heart.[211] Clearly, only the first point is truly independent of the consideration of the nutritive and perceptive systems as we have already met them in *JSVM*. For the generality of the account to be preserved, these arguments must apply to the analogue of the heart in bloodless animals and the central part of plants. The whole course of the argument seems to be shaped around a view of the heart as the central organ of sanguineous animals. As suggested above, with this assumption as a guiding principle, the search for a single central organ, and its location in the centre of living things, is more easily comprehensible.[212]

Aristotle was very impressed by the fact that the heart in the embryo in an opened bird's egg can be seen to beat at a very early stage.[213] So confident is Aristotle of the analogy between different living things that he is prepared to transfer this result to those cases where observation is not possible. The role of the heart as the seat of the nutritive and perceptive souls is confirmed by its role in nourishing, performing the final step in working up the nutrition and transmitting it to the rest of the body through the blood vessels. All the other parts involved act for the sake of the heart, i.e. its performance of the final steps in nutrition. We are referred to the treatment in *de Partibus Animalium* for a treatment of the systematic connection between the heart and the rest of the body. The main point is that the heart is the principle or source of the blood vessels, so that blood apparently emanates from the heart to the rest of the body. That the heart is the source of nutrition also explains why it is so important that Aristotle thought it came to be first: it can act as a principle causing change, i.e. growth in the rest of the body.[214]

The connection between the heart and sensation raises greater difficulties: apparently the blood vessels provide a means for transmitting sensations to the heart. Caution is in place here; Aristotle nowhere says that blood is responsible for the transmission of perception.[215] The connection which he finds between peripheral perceptive parts and the blood system may be explicable in terms of the need for nutrition in the perceptive parts.[216] Perhaps one should admit that Aristotle did not know and knew he did not know how the transmission was effected. This might be one message from the passage where he *arrives at the connection by argument*.[217] If he had had a definite view of how this worked, he would not have had to conclude that there is some such connection by analogy with touch and taste.

> For as regards nutriment the offices of the other organs are ancillary to the office of the heart and that must be the dominating organ in which the final result is achieved rather than any of these that are subservient to merely preparatory processes, as a physician who restores health is above the subordinates who supply him with material.[218]

Ogle's remark makes use of Aristotle's own comparison with a doctor (469a9) and well describes the role of the central organ: it is the top of the hierarchy of parts, which all contribute to the one end. It is this that makes this organ crucial to the life of the living thing. To say that the final food is in the heart means that that is where it is produced. The heart performs the final stage in the process of nutrition: it makes blood. The other parts – mouth, gut, liver, spleen – function for the sake of the heat of the heart.

They prepare the food for the heart.[219] Taken together, the concerted operation of these organs is the activity of the nutritive soul. Thus all parts are involved in this activity, insofar as they contribute to it and profit from it, while one part, because of the hierarchy of functions operative in the activity, is the principle of nutrition, i.e. the seat of the soul. A similar line of reasoning can be applied to perception and sensation.[220]

What is the relation between the soul as an actuality and the dominant organ? One might think that the soul being actuality in *de Anima* makes it necessary to have a single dominant organ, such as we find in *Parva Naturalia*.[221] Yet an actuality does not make such an organ necessary. For some living things do not have it, namely plants; and apparently its importance in insects is not as great as it is in sanguineous animals, if parts of insects without a heart-analogue can survive. These facts suggest that it is a question of the development of higher faculties that the heart makes possible. Superior organisation makes it necessary. But even if the soul is localised, it remains the soul of the whole body since the specialisation that makes the central organ the seat of the soul also provides it with systematic relations to the rest of the body.[222]

In this section we have seen Aristotle use nutrition and perception to locate the part with a central capacity for these functions; in this way the soul serves as guide to the enquiry. In the simplest living things the 'seat' of the soul is no more than the meeting point of up and down. In more developed things, Aristotle can point to an organ in the chest. The lack of a functionally specific seat in lower beings is clearly connected to their greater capacity to survive vivisection. The greatest problems were presented by his attempts to argue that both of the souls of animals, nutritive and perceptive, are in the same part. This seems to rest on assumptions about the unity of the life and soul of living things, particularly of highly developed living things. There is good reason to think that Aristotle is being guided by his interpretation of developed living things throughout the discussion of the central organ. It is also plausible that other factors lie behind his unifying perceptive and nutritive principles, based on their matter.

Now that we have located the soul in its body, we can turn to a fuller consideration of that body. Since we have already made use of the idea of an organ, even if in a vague way, we must make good the omission of a discussion of uniform (homoiomerous) parts, such as flesh and bone, situated between elements and organs in complexity. Some of the knottiest problems in Aristotle's natural science lie at this level.

4

Body

4.1 Balanced capacities in Aristotelian mixtures

We have already met the simplest entities in Aristotle's world, capacities for change, which in elements and things composed of elements are responsible for the perishing, or more generally change, of these things. It is now time to look at the conception of body more closely: how is it that body, especially living body, is stable? What limits its stability? Put slightly differently, how are the capacities of the body related to the activity that is the soul? In this chapter we will look at the way in which balanced capacities for change are involved in and supported by the activity of nutrition.[1] As a first step in understanding how living bodies are connected to their ingredients, we must understand Aristotle's idea of mixtures (*mixis*),[2] which is how he understands those uniform parts of bodies that make up organs. This idea has been found deeply unhelpful, precisely because the ingredients seem to lose their natures on entering a mixture. I shall attempt a very limited defence of Aristotle's conception, such that the capacities *LBV* 2 has introduced (above §3.4) are present in mixtures. In other words, I shall try to show that capacities as such, i.e. not as actualities, exist in Aristotle's mixtures. This account will not dispose of many problems inherent in this conception, but it is a step towards showing how Aristotle can think the way he does about living things and the cause of their perishing.

In *de Partibus Animalium*, Aristotle describes three levels of complexity in the composition of living things:

(1) Elements – fire, air, water and earth. Each is a combination of two of the four primal capacities wet-dry, hot-cold.
(2) Homoiomerous parts. These are parts where the parts bear the same name as the whole: a piece of some flesh is as much flesh as is the whole.
(3) Anhomoiomerous parts. These, what we call organs, are composed of homoiomerous parts (flesh etc.) and perform the functions necessary to the survival of the whole.[3]

One question, then, is how these levels are connected with one another;

part of the answer Aristotle gives is that the simpler level explains the more complex.[4] But this is not the whole story, insofar as the simpler makes a contribution to the higher level. The main problem here lies in the connection between the elements and the homoiomerous parts, which forms part of Aristotle's theory of mixture (*mixis*), described in *de Generatione et Corruptione* I 10. He distinguishes it from synthesis (*synthesis*). The latter is what happens, for instance, when we combine barley and wheat grains. The grains are not altered in the combination. Aristotle begins his account of mixtures by referring to those who do not think them possible for the following reason (327a34-b6): if the ingredients still remain and are not altered, then they are not mixed, but are in the same state. But if one of the ingredients perishes, then they have not been mixed, but one ingredient exists, and the other does not. Nor have they been mixed if both perish: non-existent things cannot be in a mixture. Since these are the only possibilities, mixing the ingredients into a mixture is not possible. Clearly, one way of avoiding this problem is to show that there is a third possibility open to ingredients on entering into a mixture besides perishing and coming to be. And this is the path Aristotle chooses, by pointing to his distinction between being in capacity (*dynamei*, often translated 'potentially') and being in actuality (*energeiai*)[5] (327b22). This allows him to steer a course midway between the ingredients' ceasing to exist in mixtures and being preserved without alteration: they can continue to exist in capacity and cease to exist in actuality: they are not (completely) destroyed in the mixing: 'Their capacity is preserved' (327b30-1). This ensures the difference between mixture on the one hand and alteration and generation on the other. In alteration, the actuality persists through the alteration, e.g. when a man (an actuality) becomes white, there is a white man, whereas in mixtures the actuality does not remain.[6] There is no actual fire in mixtures: this is part of what it means to say that their capacity is preserved: *only* their capacity, and not their actuality, is present. In generation, some new thing comes about, without preserving the capacity of what went before: the dry becomes wet – and is not present any more, for example when fire becomes air.[7] This is a transformation in which matter is described as one way of being 'in capacity': fire is air in capacity. This is to be distinguished from the persistence in capacity of ingredients in mixtures. For when fire becomes air, its capacity is lost: air no longer has the capacity to burn something; it has a completely new capacity.[8] In contrast, when warm and cold things like fire and water are *mixed* together, in Aristotle's technical sense, their capacities are preserved.[9] There is a hot, wet mixture. This is something new, different from its ingredients. Thus 'capacity' here is not merely the capacity to undergo a change; the capacity is preserved through the change of being mixed. Aristotle rejects the idea that 'ultimate particles' of the ingredient should be juxtaposed in a mixture.[10] For such juxtaposition would leave us with what he calls a 'synthesis', i.e. a mere combination of unaltered ingredients. If we recall the problem which served as the starting point for the description of mixtures, we can see why such particles are unacceptable: they would remain unchanged even in the mixture, which was one of the reasons for thinking mixture not possible. In effect, such particles would be like barley and wheat grains in a combination: actualities unaffected by their combination. Mixtures, however, form between things that can act on and be acted on by one another (328a18). They are thus reciprocally active and passive. For example, there is no mixture between health and the bodies it is in because health is not altered by coming to be in a body. This example is clearly parallel to those in *LBV* 2 (see above §3.4): health can be lost without the body also being destroyed. That is not the kind of unity we are looking for in living things.

For mixtures to exist, it is not enough for the capacity of their ingredients to continue to exist: there is another condition which we can approach by considering the following problem. When there is too much of one ingredient, a mixture does not come about: the lesser ingredient is simply swallowed up in the greater one. A drop of wine in ten thousand measures of water is not a mixture, because all the wine does is to augment the water. This follows from the fact that they act on one another: one ingredient simply wins. The answer to this is that there must be some *balance* between the ingredients which have to act on one another, if a mixture is to ensue:

> When there is a balance between the capacities, then each changes from its own nature towards the predominant ingredient, without, however, becoming the other but something between the two with common properties.[11]

In making a mixture, it only makes sense to speak of a *balance between capacities* in the mixture, if the capacity is not lost. You cannot have a balance between things that have ceased to exist. There is, however, a genuine change in the making of a mixture, since in the mixture the capacities temper one another, which they did not before they were mixed. They act on one another: there is, as it were, a balance of powers: the heat is tempered by the cold, the dry by the wet.[12] Thus there is something in between the various qualities and this something is the new thing that comes about through this mixture. Aristotle develops this idea of balance between capacities in mixtures in *de Generatione et Corruptione* II 7, by putting forward the idea that a mixture such as flesh consists in capacities – hot-cold, wet-dry – in certain proportions (*logos*) to one another. Saying what flesh is would then be to specify the capacities present in it, and what proportions to one another they are in. There may be several ways in which there can be a balance between the capacities in a mixture; each will be characteristic of a certain mixture, e.g. of flesh or bone.

We have already noted that mixing only happens between things that can act on one another, like hot and cold. Only if they act on one another can they temper one another, and so form a balance.[13] Hot and cold do not destroy one another on entering a mixture, so that one would just have hot or cold: they are in such a proportion to one another that they preserve a balance.[14] When the capacities are mixed, hot and cold are relative to one another: things can be more and less hot. Thus flesh might represent one intermediary between hot and cold, brain another.[15] In the one case there is more heat, in the other more cold; but both are situated on the range between hot and cold.[16] 'Balance' does not mean some mathematical centre between determinate ends: balance represents a range between the two extremes.[17] If there is no intermediate, the contraries destroy one another. One of them predominates to the detriment of the other. In that case, there is no mixture. As we have seen, this was one of the possibilities that the account of mixtures was designed to avoid: the complete perishing of the ingredients, since that would give us a case of coming to be, and not a mixture.

The difficulty with Aristotle's conception of continuous matter is acute here: how are we to understand the presence of these capacities in a mixture, if they are not in some way locally separate? For if the capacities in a mixture are locally separate, we lose the homogeneity of the mixture: if this bit is hot and that bit is cold, then the whole is not uniformly such and such a proportion between hot and cold. And that is the definition of the homoiomerous parts such as flesh that we are trying to explain: any part is like any other. In a sense, this is the point of this conception of mixture. The capacities of the mixture are present in the *whole*; if they were localised, there would be no homogeneity. These points form part of Sarah Waterlow's criticism of Aristotle's conception of mixture.[18] It will be useful to consider the points she makes before attempting our partial defence of Aristotle. In modern conceptions of chemical combination there is an identity between the atoms in a compound and those not in a compound. In contrast, it is impossible to divide an Aristotelian mixture in such a way as to arrive at something with different properties from the whole: the whole cannot be viewed as a system of component structures. Homoiomerous parts are simply homogeneous. For Aristotle, simple elements are essentially characterised by their locomotion in a given direction. This fact bars his way to the idea of a compound. For any compound of such elements would be very unstable, and only held together by a miracle. The solution Aristotle adopts for this problem is to say that the elements are only potentially present, as we have seen above. Waterlow understands this to mean only that the ingredients can be separated out again. The effect of all this for Aristotle's biology is that for Aristotle there can be no question of explaining the structure and behaviour of organisms and organs by reference to the properties of their simple components.[19]

She concludes that ingredients must either *totally* lay aside their nature on entering into a compound or else modify it in such a way that they fulfil the needs of the whole. There is much that is unanswerable in this criticism; but, while there are obviously great problems in understanding Aristotle's continuous conception of matter, and this leads to special problems in the field of compounds, I think there is rather more to be said in favour of his approach than Waterlow allows. It may be impossible to locate structures in homoiomerous parts, but one can say something about the composition of such mixtures, namely what capacities are present: there is some sense of structure in the mixture, because the capacities are in proportion to one another, and these proportions define the mixture. In this way, the ingredients in the mixture do not have to be modified arbitrarily to fulfil the needs of the whole; the nature of the mixture depends on the capacities present in it.

Because the capacities are in proportion to one another we can see the solution to another problem raised by Waterlow: the instability of mixtures due to the elements tending in certain directions. Waterlow suggests that Aristotle's way out of this is to say that the elements are only potentially present, with the implication that this means they no longer tend upwards (fire) and downwards (earth). Only then can Aristotle talk of mixtures at all. On our reading of GC I 10 and II 7, the elements are not present in the mixtures as actualities, but their capacities are: to move upwards, downwards, moisten things, make things hot. In these capacities, we can see why mixtures cohere for Aristotle: the hot does not fly up and the cold fall, nor the wet flow away, nor the dry crumble *because* they are mixed. To some extent, they cancel one another out: it is reasonable that if something light is to be prevented from rising, something heavy is needed to stop it. As is required by the account of mixture given above, they must act on one another if they are to mix. Similarly with something wet: if it is not to flow away, something solid is necessary to give it shape. And so on with the other qualities: indeed it seems reasonable to think that it is this that is the reason for thinking of matter in terms of capacity, namely to act on other matter. In a mixture such capacities do not neutralise one another entirely.[20] For example, a living body is still hot and wet, although it may also contain dry and cold stuff.

In *GC* II 8 Aristotle gives reasons for mixed bodies being composed out of all four elements. That he should trouble to do this speaks against regarding ingredients as *completely* denatured in mixtures. Mixed bodies (*mikta*) contain earth because they have their place in its place. They must have water in them to bind the earth: if something dries out it no longer coheres. Fire and air are necessary in order to balance the other elements to which they are contrary. This line of argument underscores the importance not only of balance to the mixed bodies, but also of the fact that the ingredients have to act on, and undergo change through one another. Waterlow presents us with a dilemma: either ingredients lay aside their nature entirely on entering a mixture, or they modify it so as to fit in with the function of the whole.[21] It should be clear now that indeed both of these points are correct: the actuality of the ingredient as fire is laid aside; and its capacity as hot is modified (tempered) by the other ingredients. But it is not so clear that this is as harmful for Aristotle as Waterlow thinks. For the preserved capacity presents a connection with the ingredient – heat from fire, wet from water. If the nature of ingredients is only actuality, then they lose it; but if their nature includes capacity, then this is preserved in mixing. There may be no actual fire in a mixture such as flesh, but there is a capacity to heat things. Talk of balance should not hide the fact that mixtures and the uniform parts of living things in particular are not static balances, but balances in which one capacity, heat, is able to master and keep mastered the other capacities. We might call such a balance a dynamic equilibrium. After all, it is a balance between capacities, *dynameis*.[22]

A good analogy for Aristotle's conception may be seen in a sauce.[23] Suppose it is flavoured with a variety of herbs. In order to comprise *this* sauce the herbs must be in such and such a proportion to one another. When one tastes such a sauce, it is tasted as a whole, without pointing to the individual particles of flavour. But the recipe prescribes the ingredients and their proportions. The flavours may not be destroyed on entering the mixture, since they are still there, although not isolated, and together constitute the flavour of the whole. An individual flavour, however, may not be sensed in its own right, but only as a component of the composite flavour. This composite flavour comes about through the mutual influence of the flavours on one another. Any one flavour may not predominate in such a way as to mask the others completely. If it did, one would say, not that the sauce is flavoured by the different spices and herbs, but just by this one. Similarly with a mixture. If one ingredient predominates to the destruction of the others, it is no longer a mixture, but just that ingredient.

There are two aspects of Aristotle's conception of mixtures which I hope are now more secure than they were: the existence of capacities in mixtures, and the consequent possibility that these capacities are in some kind of equilibrium. These are the aspects of his theory that are necessary to understand functioning and its failure, perishing, in complex things. It must be admitted that problems remain with the localisation of capacities: just how is it that capacities are localised by Aristotle? In general he works back from an activity, change or actuality, for which the capacity is a capacity. On the most basic level, he starts from the way in which things affect us, that is say cause changes in us, as hot, cold, wet and dry things. The simple qualities in his world are capacities to affect us and also, but not in the same way, each other. This affection takes place because the capacities are capacities for change.

4.2 Living things and their surroundings

The living things we are talking of are finite or limited, and part of what this means is that they live in surroundings. But how are they related to their surroundings? And how is their local limitation, as we may call it, related to other forms of limitation, for example that they live so long and no longer?

A basic question to be asked of finite things is: why must they perish? Aristotle has an argument for this in *LBV* 3.[24] The conclusion of chapter 2 was that there is no perishing of the soul separate from the capacities that destroy one another. The capacity to perish is based on the contrariety of the capacities in the bodies. Now we need to look at contraries and ask: *must* they destroy one another? The chapter starts with fire in the upper regions as a test case to see if anything that has capacities with contraries might be thought to escape those capacities contrary to it. He continues (465b3-8):

> For things present in contraries perish accidentally, through the perishing of the contraries. For contraries are done away with by one another. But none of the contraries in substances perishes by accident, since substance is not said of anything. So that it is impossible for something to perish, if and where there is no contrary to it.

We begin with the perishing of accidents: accidents, non-substances, can perish because what they are in perishes: our knowledge perishes when we die, and we perish through the action of contraries on one another.[25] The contraries do away with one another and, in so doing, do away with whatever is accidentally present in them. That is, something in contraries perishes accidentally. Problems arise with the perishing of substances, that is primarily living things: no contrary in a substance perishes accidentally, since substance is not said of anything.[26] In this passage, substances are introduced almost as an aside, although they are the perishing things we are mainly interested in. That is to say, perishing is to be explained using contraries, and since substances are not contraries, their connection with contraries must be established. But this passage leaves us wondering just how the contraries are related to the substance: the substance is not said of the hot-wet stuff. One way of understanding this is as follows. You cannot simply say: hot wet is a cow (we need to know what makes the hot wet stuff a cow) but rather: a cow is hot and wet. It would appear that the contrary involved in the change that brings perishing with it, follows from the essence of the substance. Contraries are not merely present in a substance, as paleness may be present in Socrates; contraries here make a contribution to the substance. While a cow is not the contrary of anything, being hot and wet follows from being a cow. That is to say, such attributes will form part of the explanation of the subject. But given the need for balance between the capacities for change in a living thing, if its activity is to be preserved, it is clear how the move from contraries to cow can happen, without wanting to say that the cow is an accident of the matter it has made its own. For if the cow is, formally speaking, an actuality involving hot and wet, then the actuality is the subject of the hot and wet, in that the actuality is the cause of the cow being hot and wet.

The general conclusion of the passage quoted above is that nothing perishes without the presence of its contrary. The first half of *LBV* 3 establishes the possibility that something, although it has a contrary, will be non-perishable where that contrary is not. Hence it will be possible for it not to perish. The second half of the chapter is designed to close this gap.[27] The basis of this proof consists in the claim that everything that has matter must have a contrary.[28] Aristotle does not think he needs to reason for the perishability of contraries because they do away with each other.[29] Since qualities such as hot are in matter, they are always given with the capacity of change to their contrary. Thus the contrary is there, wherever heat is, but only in capacity. All that one then needs for this to be actualised is a moving cause to actualise this capacity. 'Heat ... can be everywhere [in a thing], but not the whole thing can be hot ....'[30] Heat is in all the matter underlying fire, like heat throughout hot water: it is in a subject but is one of a pair of contraries. Thus the subject comes with contraries. The heat is spread throughout the matter, like heat in hot water.[31] That is to say, it is not just this bit that is hot and that bit that is cold, but the whole thing is hot. Nonetheless, the whole thing is not heat; for it is a composite of heat and matter. Whatever has matter has a contrary because if things can be hot, they cannot be simply heat. You cannot attribute separate existence to qualities, which must be attributes of substance, but heat, which is in a subject, has its contrary, i.e. the contrary state of the subject. An animal which is warm can cool.[32] An animal can be hot, but heat cannot be all that the animal is, since heat is not capable of existing on its own: it must be in something.

This leaves open the question of what makes the hot thing cool. This hot thing only has the contrary cold insofar as it can be cold.[33] Thus there must then be something to make it cool:

If, whenever the active and the passive are together, the one always acts (*poiein*) and the other undergoes (*paschein*), then it is impossible for there to be no change.[34]

When one capacity for change, e.g. for heating, approaches another, e.g. for being heated, change is necessary: and then heating takes place. The water in the kettle has to be close to the fire to be warmed and thus actualise its capacity for being warmed. This applies to the effects of cooling, heating, wetting and drying out that occur between the elemental qualities hot, cold, wet and dry. The further sections in the chapter are concerned with where this active cause is situated relative to the body on which it acts. It can be internal to the thing that decays. So residues in living things are discussed: they are contrary to the heat, and hence cause cooling off. Residues (*perittômata*)[35] are produced by the heat of the living thing. Some are by-products from the production of useful parts such as tissues; others are useful in themselves. As well as being in the body concerned, the contrary can also be external: warm bodies can be in cold surroundings.

Change is therefore only necessary if a moving cause is present: perishing is conditional on the presence of a contrary outside or inside the thing.[36] Aristotle is then prepared to *posit* (*hypothesthai*) the presence of contraries inside something if the environment is not enough to cause the process.[37] This makes good sense if we are concerned, as we are, with living things. They are mixed bodies, as was explained in *LBV* 2, and, given that such bodies do perish, perishing occurs through the effect of contrary capacities on one another. But in what way are we talking of living things? As composite things containing contraries, which, when appropriately together, always (i.e. necessarily) act on one another; or as forms for which matter is necessary (*if* there is to be a cow, then such and such matter is necessary).[38] This second level of necessity, so-called hypothetical necessity, is not at play here, for the following reason. *LBV* 3 also speaks of a greater fire using the nutrition (*trophê*) of the lesser one.[39] If fire feeds, produces and expels waste, then animals appear in *LBV* 3[40] as only one kind of substance (*ousia*) in which there are contraries. Nonetheless, the consequences of the argument are drawn in terms of living things: the environment (*periechon*) can influence life-span favourably, but not so that the living thing is everlasting.[41] This can be explained by noting that it is stated completely generally that it is impossible for something with matter to have no contrary.[42] The argument in chapter 3 is about process generally, the conclusion is about longevity. It is revealing that there can be a dispute whether living things are meant or not. Decay of living things must first be treated like any other process, with a moving cause and change between contraries, before one goes into the full picture of how decay fits into the functioning of the living being.

Aristotle uses this confirmation of coming to be and passing away in things with matter to explain one of the problems mentioned in *LBV* 1, namely why kinds (*genê*) of men in hotter places are longer lived. The environment either increases or decreases the length of life within their natural limits. Because the presence of the actual contrary can be internal or external (i.e. the environment), even though the form (*eidos*) of an animal is the same in both localities, the environment can make it behave differently: it can encourage or discourage coming to be and passing away.[43] This latitude in the length of life might depend on the idea that, while a constitution allows some variation (e.g. in size), this variation is not arbitrary. This is one explanation of the variation in longevity between members of a kind, rather than between kinds.[44]

For Aristotle all change is dependent on locomotion – the locomotion of the heavenly spheres.[45] Ageing, conceived of as a change from hot and wet to cold and dry, depends on the locomotion of the sun and hence on the seasons. The changes in the seasons change the hot and cold in environment; and the hot and the cold in the environment affect the living thing.[46] Cycles in nature are said to be an imitation of celestial cycles[47] – thus life-spans, one might think, are a consequence of heavenly revolutions,[48] giving a global theory of longevity. The reason for the periodicity of lives lies in the motion of the sun along the ecliptic: its approaches to earth bring about growth, through the increased heat, and its recession, decay:

> Hence, too, the times – i.e. the lives – of the several kinds of things have a number by which they are distinguished; for there is an order of all things, and every time (i.e. every life) is measured by a period. Not all of them, however, are measured by the same period, but some by a smaller and others by a greater one; for to some of them the period, which is their measure, is a year, while to some it is longer and to others shorter.[49]

The 'periods' by which things are measured are not the units measuring the life-spans, but rather the results of measuring the life of each species.[50] It is more difficult to know what to make of the order (*taxis*) of all things. To understand it as a kind of celestial timetable,[51] from which one might read off life expectancies, is undesirable, not merely because such a reading would conflict with my reading of Aristotle's theory, but because he is generally a minimalist when it comes to asserting the order holding between parts of the universe;[52] and having such a plan might imply a very strong order controlling all things, rather than the nature of things controlling them. Nonetheless, we are here faced with an ordering greater than that of the individual kinds of things; indeed such an order is implied by the idea that living things fit their environments, drawing sustenance from them. And the present theory of periods can well be understood as an extension of this fit. In a regular context, unruly lives are unnecessarily tough. There is an order of all things, not because this order is prescribed, but because the natures of the things concerned involve them in an order. Looked at from the perspective of *de Generatione et Corruptione* II, where we are concerned with the general cause of coming to be and passing away, it may look as though transient things had little say in their life-span – in short that the heavens provide a timing device ruthlessly determining the ends of things, in the relevant sense of 'ends'. But there is a gap in the theory of *GC* II 10, which presents a window of opportunity for the activity of living things. And that fits comfortably with the rest of Aristotle's theory as presented so far in this section.

Aristotle appeals to what we see happening around us to confirm his theory – presumably the seasonal aspects of vegetation, annual and diurnal insects. The simple correspondence between warmth and growth, decay and cold cannot disguise the fact that many living things such as ourselves, if really affected by the ambient warmth in our longevity, are so in a way that does not obviously obey this simple pattern. And such a theory seems to emerge at certain points in Aristotle; cold can be healthy, heat can destroy the natural heat of living things.[53] Alongside this complication, an obvious added aspect of the theory lies in living things that, being highly developed, are very hot, and are able to preserve this heat themselves, at least up to a point. Part of what this means is that they are not so much at the mercy of the vagaries of their surroundings – as plants are to an extreme degree. Rather, their growth and health (functional balance) are affected by their surroundings, but not immediately terminally.[54] It is thus possible to locate living activity in the theory of *GC* II 10, although it is not mentioned there as such.

The theory of *GC* does offer a global justification for the idea of a life-cycle in general, as opposed to explaining why this kind of thing lives this long.[55] For cycles are a way of ensuring existence under material conditions: the changes in matter are ordered[56] in such a way as to make coming to be uninterrupted (*endelechê*), and so as close to being as possible for such changing things. Since these things *exist* in this way, they fulfil the requirement that being is better than not being. Living things do not return in a cycle, such as that of the elements, continually turning into one another, but their succession is that of one generation after another. This form of imitating the heavenly cycles in a modest way provides an external motivation for Aristotle to hold onto the idea of cycles. So in this sense one might well think of the heavens imposing their pattern on living things, even though this pattern allows them specific variety, and the internal explanation of their life-cycles.

Elsewhere Aristotle refers to common sayings that time wears things away, makes us forget things, and makes us grow old; and says that this follows from his definition of time as the number of change in respect of the before and after, in that change removes what is there.[57] Even when we know that Aristotle does not think that time itself does anything, we are left wondering how time manages to do away with things and leave nothing in their place; Aristotle admits that it is also the cause of being 'but only accidentally': how can one explain this bias in his view of time? Edward Hussey has suggested that the 'action of time', as he calls it, primarily affects living things, which, even when not worked on from outside, decay because of their internal structure.[58] Such beings are sufficiently closed off from their surroundings for it to make sense to say that some changes (such as sickness) come from outside, and others (such as ageing) do not. Such degenerative change is strictly due to no one cause, and so is ascribed using a metaphor, to time.

There are a variety of problems with this reading of the action of time, which I wish to sketch, although I can offer no alternative reading. Foremost among the problems is perhaps that we are asked to believe that Aristotle did not think that non-living things simply decay in the course of time. But there are also problems with the application to living things – particularly that there is no specific cause for such decay. For Aristotle this would imply that there can be no scientific knowledge of such decay; and yet we seem to get at least some gestures towards such knowledge in the *Parva Naturalia*. One more problem can be mentioned. Leaving aside the question what kind of a possibility it might be for something not to be worked on from outside, it is clearly true that Aristotle does believe in natural ageing, and old age is one of the things he names as an effect of time. But equally clearly he believes in natural maturing, and that is not, by definition, a degenerative process; and it is not clear that a thing when maturing is any more or less isolated from change from outside than an ageing one.

Being in time is an essential property, and entails that the thing concerned can be measured essentially: lasting this long rather than that long is inherent to the thing.[59] Now we have seen to what extent Aristotle can accommodate living things' surroundings as a necessary element in their perishability, which however does not deprive them of the capacity to live their natural span. Indeed, because finite beings are dependent on heat from outside for their activity, the background regularity in their surroundings forms a basis against which their specific profile can exist.[60] Particularly here, when we are talking of measurement, the lack of numbers in Aristotle's account is striking, and his restraint is maintained even when he is talking about the correlation between length of gestation and life-span in *GA* IV 10.[61] The correlation lies in the fact that both 'tend to be measured naturally by periods, i.e. day, night, month, year and times measured by these'.[62] The basis for this theory, as the reader of *GC* II 10 would expect, is heatings and coolings caused by the heavenly bodies. Up to a certain 'balance' (*symmetria*) these produce generation, and afterwards perishing.[63] The emphasis of this passage is clearly on the external causation of periodicity, and indeed the scale of nature seems to lie behind the theory: it is reasonable that the periods of things with less authority (*akurotera*) 'follow' those with more.[64] Yet not only is the thesis here imprecise ('follow' might mean anything from existential dependence to a loose correlation) but a let out is offered by Aristotle's concept of matter: the tendency of things to be measured by heavenly periods is attenuated by the indeterminacy of matter and the plurality of principles that impede natural generation and perishing.[65] As we saw in *GC* II 10, Aristotle has a very commodious notion of order, which remains despite exceptions. It is, as there, closely connected to the fact that change can be piled on change without affecting the basis of order, namely the heavenly revolutions; but then nor do the heavenly motions provide a serious threat to the possibility that interior organisation is the factor that, given the heavenly motions, is what specifically determines how long something lives.[66]

In this section we have considered some of the relations of living things to their surroundings, in particular the extent to which their life-spans are *measured* by their surroundings, more exactly by the movement of the sun along the ecliptic. We have seen that this relation of the capacity for change inside to capacities for change outside does not dethrone the activity of living things from their central place in explaining a life-span that can vary within certain limits by living in different surroundings.

4.3 The explanation of life-span: the hot and the wet

As the first step towards seeing how the capacities for change in living things explain aspects of their living behaviour, we shall continue the exposition of *LBV*. In chapter 2 of that work, as we have seen (see above §3.4), Aristotle decides that it is body rather than soul that is responsible for making living things easily perishable. But that does not tell us what it is about living things that decides their longevity; as an approach to this question, in chapter 4 Aristotle goes through various contrary groupings of living beings. Thus we are comparing the classes of living beings from which we started in chapter 1.[67] The chapter divides roughly into two. The first half (466a1-9) dismisses any pairs of classes – large/small, plant/animals, sanguineous/non-sanguineous, land animals/aquatic animals – on the grounds that none of them corresponds to long-lived and short-lived: there are always exceptions which prevent any pair from matching the division between the long-and the short-lived. None of these differentiae is simply the cause of longevity.[68]

In the second part of the chapter (466a9-16), Aristotle makes broad (*holôs*) distinctions, i.e. ones that hold despite the exceptions. He maintains that plants are longer-lived than animals, sanguineous animals longer-lived than bloodless ones. One of his major tasks in the coming chapters will be to reconcile the general distinctions with the particular exceptions. He continues here by combining the differentiae 'sanguineous' with 'land animals' and says that such animals are longest-lived. As Balme remarks on Aristotle's practice of differentiation, 'he looked for causes by examining combinations of differentiae'.[69] In our passage, Aristotle has isolated indications, but has not yet named the cause, which is what we are after in the work as a whole. He then adds size to their significant attributes, without giving a reason for doing so.[70]

We should note the preliminary status of such differentiae: the clues they offer will be taken up and made into an explanatory theory in chapters 5 and 6. In chapter 5, we move from the results of the preceding chapter to find the cause of long and short life:[71] Aristotle takes a new piece of evidence not mentioned in chapter 4, namely:

> We must understand that animals are by nature moist and hot and living is such, and old age and the dead are cold and dry. For they appear so.[72]

The first thing to note about the remark about the old and the dead being cold and dry is its generality: it cuts across the differentiations of the previous chapter to make a claim about old age in general. The evidence he gives for this fact is that living things 'appear' so. Whether this is an 'appearance' in the sense of a perception, or in the sense of a reputable opinion, is not clear. The fact that the appearance involved is a pair of perceptible qualities would suggest that perception is involved. It might be a reputable opinion because the perception that living things are hot and wet, while the old and the dead cold and dry, is so obvious.

An added factor should be taken into account: this statement occurs within the context of Aristotle's enquiry into living things. In particular, their heat plays a crucial role, as we have seen in the central case of nutrition. It would be hasty to think that heat is introduced with no allusion to the role it plays in living things. By the time one arrives at *LBV* 5 one should, at the very least, have read *de Anima* II 4 which makes clear that living things have to be hot and wet to feed,[73] and, in the case of animals, to perceive also. Not to mention the many other places where heat is vital in Aristotle's account of living things and their functions. Rather than just perceiving the old, we are perceiving them bearing in mind some knowledge about homoiomerous parts and their functions in the whole composite of body and soul. That means that this piece of 'evidence' is to be seen in the context of an investigation of vital function;[74] it is, as it were, an educated appearance – the way things appear to a trained observer. This evidence is connected with the differentiations of chapter 4 via the hot-wetness of large sanguineous land animals, which are the animals the differentiations arrived at as being longest-lived. That they are the longest-lived suggests that the quality which is most characteristic of them is connected to longevity in general. Vital heat alone does not explain such differences.[75] The matter of living things is said to be the four basic capacities, hot, cold, wet and dry, clearly a development of chapter 2's insistence on the responsibility of the contrary capacities for perishing in living things.[76] There is some more discussion of the matter involved:

> It is necessary for animals to dry out as they age. Thus the wetness [sc. in them] may not be easily dried; and so, fatty things do not rot. The cause of this is that they contain air and air is related to other things as fire, which does not become corrupt.[77]

These lines are a beautiful example of the two faces of Aristotelian explanation, despite the unclear connection between fat and air. On the one hand, the requirements of proper functioning into old age make it necessary for animals' wetness to be hard to get rid of; on the other hand, the persistence of their wetness is laid at the door of the behaviour of air, which has gone into the make-up of their fat, here to be taken as a moisture hard to dry out, since, as Aristotle says, it is heat in some animals, and in others there is some other juice. Note that it is the capacity, the heat, which does the explaining. Later in the chapter Aristotle cites the fact that bloodless animals have no fat as one of the reasons that they do not live so long.[78]

Among the general observations in chapter 4 was the one that larger animals live longer.[79] While no reason was given before, it is now grounded in the fact that they have more wetness than smaller animals. Aristotle connects this implicitly with the previous point about the resistance of the wetness to drying up, when he goes on to say that it is not just a question of quantity, but also of the quality: the wetness has to be hot as well as plentiful. Using a combination of these two criteria, the quantity and the quality (hot-cold) of wetness, he can thus explain an exception to the generalisation set up in chapter 4, namely that horses, although larger than men, are shorter-lived: they may have more wetness but it is not as hot.[80] He sets up a proportion between the wetness of the one and the heat of the other: an animal is longer-lived if the amount less it has of wetness is compensated for by the amount more it has of heat.[81] The heat must be in proportion to the moisture in the organism. If it is too little, the moist element is too easily frozen and the hot itself quickly done away with. But, on the other hand, if heat is there in too great a quantity, the moisture is quickly dried.[82] This is again the all-important thought that there are proportions of the different capacities in different things. The difference in longevity between horses and humans is to be accounted for by the difference in the proportions of the capacities in them. This is as near as Aristotle comes in *LBV* to talking about the actual lengths of lives; but even here, the point is a comparative one. The groupings of chapter 4 have been picked up. Size and being land animals have been played off against one another using the quality and quantity of wetness.

Saying that the hot and wet are long-lived suggests that it is by the loss of hot and wet matter that things lose their longevity, so becoming dry and cool. And this is the message of the next section (466b4-16): residue (*perittôma*) such as seed can remove hot wet stuff. Residues arise from the use of nutrition in which heat is involved. The food is worked on, 'concocted', in a series of stages, dividing the useful from the rest, until the stuff for which there is no use is passed out of the body. The consumption of wet-hot in each case fits into the economy of the animal's way of life.[83] Residues presumably cause death because they are present, and have to be got rid of for functioning to continue, which places a strain on the living thing's capacity to cause the relevant kind of change. Residue is said to cause death in two ways. It does so naturally if it is contrary to the whole blend (*krasis*) of the body, and hence overturns (*anatrepein*) the blend of the body. Or residue can be opposed to one part of the body. This would seem to be a reference to sickness in a part.[84] Aristotle gives examples, such as 'salacious' sparrows that, by emitting a lot of seed, lose their connate heat and hence shorten their lives.[85]

There is an obvious and intimate connection between health and perishing: a healthy body will not be as liable to perish as a sick one,[86] and although this need not mean that all perishing is morbid, our treatment of capacities in the last section enables us to give a good account of health and sickness here. These are states of a living body, and are said by Aristotle to consist in or to be a balance (*symmetria*) of the hot and cold in the body. There are, as Robert Wardy points out in his discussion of these concepts as they appear in *Physics* VII, two ways of understanding this balance, which may be said to complement one another. The balance can obtain between the hot and the cold in the body; or between the body and its environment. Because Aristotelian mixtures are not to be characterised by the relation between 'independently subsisting distinguishable ingredients', Wardy concentrates on the relation between the living body and its surroundings. He then suggests that some feature of the body could be healthy in one environment and unhealthy in another: being fat may be good if one lives in the Arctic, less so near the equator. In this way, 'health' will be a relational predicate, because the body will possess it in relation to the environment. However, this suggestion still requires that when the healthy state of the body is reached (being *this* fat), the body has to undergo some real change; so Aristotle's view of health still requires that real changes within the body be taken into account. Balance with the environment is not enough.[87] However, now that we have shown what it is that can be in balance or harmony in the living body, namely capacities, we are in a position to supplement this account. While there may be no actual components in Aristotelian mixtures, capacities are there, and so may be in balance with one another or not. Changes in the balance between hot and cold within the body can constitute health, understood as a relational property, i.e. as constituted by relations within the thing or between it and other things, or both kinds of relation. Health depends on the relation between hot and cold in the body. A further advantage of this account is that for Aristotle, as we shall see, the environment acts on living things through being hot and cold, among other ways. The balance between hot and cold within the body may therefore simply be affected by the hot and cold outside it.

If health and sickness are in part a matter of the balance between the capacities in a living body and its environment, the same may be said of longevity.[88] We encountered this problem in chapter 1, namely how it is that 'kinds' (*genê*) of men live for different lengths of time in different localities. The change in environment does not make the nature of the animal causally irrelevant; it modifies the way it is expressed. In chapter 3 the contraries necessary to cause change are either in the environment or in the living thing, as residues. There, the environment is said to be able to make living things longer- or shorter-lived.[89] This implies not that there is no specific age (an age range characteristic of the species), but that there is a degree of variation. How a body is acted on from outside is still dependent on what kind of body it is. In chapter 5, Aristotle can complete his account of the environment in relation to longevity. What is affected is the wetness qualified, more or less as the case may be, as hot. For he says that animals are long-lived for the same reason that they are large in hot places: 'Hot wetness is the cause of growth and life.'[90] This means that where there is less heat, certain animals do not grow at all, and others grow to be less large. The effects of the environment are more obvious on animals which are naturally cold, such as snakes and lizards (466b19 f).

The next section touches explicitly on the reason that hot-wet is so important: animals and plants require food:

> Both plants and animals are destroyed when they do not take nourishment. For they colliquesce (*syntêkein*) themselves. For just as a lot of flame destroys a small flame by consuming its nutrition, thus in this way, natural heat, the primary digestive faculty, consumes the matter it is in.[91]

The idea is that if a living thing does not feed it consumes itself; the innate heat, i.e. primary concoctive faculty, consumes the matter it is in. This can be compared to the consumption of one flame by another[92] since the flesh the connate heat is in is also hot. The word Aristotle uses for 'consume' is *syntêkein* – a word which, in David Balme's words, means 'a morbid breaking down of tissues into noxious fluids',[93] and the point is that this breaking down can take place as a kind of burning like that of nutrition, but different in being morbid. The result of such a breaking down, colliquescence (*syntêgma*), is a secretion (*apokrisis*) produced against nature from the material that supplies growth, unlike residue. The latter is the product of natural or healthy concoction, as we have already seen. Here, the mention of the fact that natural heat is seen as the primary faculty of concoction makes clear the connection between feeding and the hot-wet quality of living things. The idea is presumably that you get thinner if you do not eat: Aristotle sees this as flesh being burnt up by natural heat.[94] Some fuel is needed to keep natural heat going, and if the matter of the body is all that is available, then it will be consumed. Obviously, the lack of mention of the nutritive soul here should not be taken to imply that it is not to be assumed in the operations described: the changes described here do not affect the soul, but the body, which is hot, serving the soul.[95]

Chapter 5 shows Aristotle using the conclusions of chapters 2 and 3 to solve the central problems of the treatise. He builds on the conclusion from chapter 2 that natural unities have no perishing apart from that of the contraries, and of chapter 3, that these contraries must destroy one another when appropriately related. Chapter 4 offered rough criteria: large size, sanguineous, land animals, to guide the search for the decisive criteria. These are then decided on at the start of chapter 5 from the way living things, aged things and dead things appear: the length of life is decided by how hot and wet living things are. That these qualities or capacities are to be seen as connected to the function of nutrition, anyway to be expected through their function elsewhere, becomes explicit in the way these capacities can be exhausted, e.g. by residues from nutrition. Aristotle thinks the student of nature can identify what it is that causes life to last as long as it does, namely hot-wet, and it is the change from this to dry-cold that constitutes ageing. The *length* of this process is affected, first by the amount of wet, and secondly by how hot this wet is. It is further affected by the kind of life in which it takes place. Using this account Aristotle then tries to clear up the puzzles he has so far encountered. It becomes clear that food is necessary to conserve the wet-hot, i.e. to prevent the natural heat of living things from consuming them.

LBV 6 picks up the comment in chapter 4 that the things that live longest are plants.[96] The differentiation between plants and animals is meant to cover all living beings that age. *Parva Naturalia* is a general study of living things. Now, with the consideration of plants, Aristotle answers the question put at the very beginning of the work, whether plants and animals have the same causes for their longevity or not.[97] Throughout the final chapter there is a three-stage comparison between plants and animals.[98] First, this comparison operates on the level of wet-dry.[99] Thus we have recourse to the principle set up in chapter 5 that hot-cold, wet-dry are the matter of bodies for animals.[100] This is now transferred to plants, which win out, again on the principles laid down in chapter five for animals, as being less watery and hence not freezing so

91

easily.[101] Exactly the same reasoning is used on the one hand for animals in cold places, and for water-animals in chapter 5, and, on the other hand, for the oiliness which accounts for the longevity of plants, and which plants have and bloodless animals do not.[102] Plants' stickiness seems to be unique to them.[103] Apparently, it makes drying out more difficult. Because of the functional role of heat, everything depends on its preservation; but there are different methods of achieving this. As we have seen, this is a central strategy of the *Parva Naturalia*: to see how one process may happen in a variety of ways.

Secondly, the reason for the longevity of trees is said to be unique to plants, having an analogue only in insects.[104] This section forms the bulk of the treatment of the longevity of plants. The fact that plants may be divided and live, while it is a striking feature of plants, is not associated with their being plants, that is with their lack of perceptive soul. This might be the reason that Aristotle is so interested in the divisibility of insects. He is taking note, in his way, of similarities between plants and animals; primitive animals are close to plants. Both must comprise hot and wet material parts to perform their vital functions of nutrition and reproduction.[105] The divisibility of plants grounds their longevity. Plants are long lasting 'since they are always in process of becoming young'.[106] What he means by this is that different parts of the plant are continually being replaced, different parts at different times perishing and coming to be. Thus over time the whole being can replace itself; this distinguishes them from sanguineous animals. The analogy of plants with insects lies in their capacity to be divided and yet live. This implies a very particular principle of unity for such entities. The difference between plants and insects is that in the latter, while they can be divided and live, the pieces do not have all the organs to survive; they have either a stomach, or a mouth, but not both, nor can they produce these organs.[107] On the other hand, plants *can* produce the necessary organs since any part of a plant is potentially stem or root, that is to say, it has the capacity to produce stem or root. Aristotle thinks that this process of regeneration resembles making cuttings. This provides the reason for an otherwise puzzling detail in the account. We can understand that a perennial might lose its leaves and stalk, and regenerate. But this regeneration is not obviously so easy starting from any other part of the plant. For how can a plant survive without roots, even if it has the other parts? Obviously, only when you make a cutting and it provides its own roots. There is a distinction between a cutting and new shoots in that a cutting is separated from the original plant.

The third comparison between plants and animals concerns the longer lives of males among animals and of those plants with big roots. Plants and animals are similar insofar as large-headed animals and large-rooted plants live longest. The comparison rests on two of Aristotle's convictions:

that roots are analogous to heads and that males are hotter than females and hence longer-lived.[108] This property of plants with large roots is grounded here by their being more 'dwarf-like' which in this context must mean 'with bigger heads'.[109] Since the brain is there to counterbalance the heat of the heart, it must be larger to counterbalance a greater heat.[110] Thus organisms with larger heads, or roots, are hotter, and hence longer-lived, than those with small ones. This must be connected to nutrition too. Large roots enable plants to feed themselves better and hence survive longer. So plants do live a long time for the same reasons that animals do: through their heat and moisture. But the ways in which these qualities of living things are preserved differ. In both cases the way of life, as well as the organisation of the living thing, must be considered. The qualities that have to be preserved are preserved in the context of the life of the plants and animals, in view of the functions that these qualities serve.

Longevity is affected by the loss of heat and wet in the whole living thing and its parts; Aristotle gives his account of such decay in *Meteorology* IV 1.[111] The change that occurs to the whole organism over the latter part of its life is viewed as a special form of the change which happens to a tissue or organ. Aristotle calls it rotting (*sêpsis*), and it is to be understood as a special case of perishing.[112] To understand this we have first to consider its contrary, coming to be. This is a continuation of the story of the constitution of mixed bodies which we told in the last section. Heat and cold are active capacities, wet and dry are passive. Their interaction causes simple generation, change and perishing.[113] Simple and natural coming to be is through these powers 'when present in the right proportions' from the matter underlying each thing, i.e. the wet and the dry: 'The hot and cold cause coming to be by controlling the matter.'[114] In the case of living things, this controlling or mastering is the work of the nutritive soul and is to be identified with concoction – heat serves the soul (see above §3.3, below §4.4). Mixtures come to be when they are concocted: this is what the hot does to the wet.[115] The contrary to this coming to be is rotting, caused by a failure of concoction. Lack of heat causes decay because of what heat does: it controls, masters the wet.[116] The control is lost when that which bounds, i.e. the heat, is controlled by that which is bounded, i.e. the wet.[117] Old age occurs when there is no longer control of the wet-dry by the hot-cold. Because of the way in which the wet and the hot are combined, destroying the heat will also in the end destroy the wet. This passage from *Meteorology* is invaluable for the emphasis on the connection between control and balance. Heat will only be in control of the wet when there is enough heat. That is to say, there must be a certain proportion between heat and wet for the heat to remain in control, thus preserving the mixture. Things get old because they are no longer under control and become destabilised. Here we have control, not just at the level of the organism as a whole over its matter, but of one 'matter' over another: even among matters there is control. The matter in question is, of course, the immediate matter of the organism.[118] It is the heat in the flesh that preserves it.

Since mixtures consist of hot, cold, dry and wet, they must be able to protect their proportions from being destroyed by external influences. This touches a point of great importance in Aristotle's account of living things: the effect of the environment, i.e. that which surrounds the contained thing:

> Rotting is a perishing of the wet thing's own natural heat through external heat. This latter heat is that of the environment.[119]

On the face of it, this might lead one to think that decay is after all a mere imposition from outside; this definition clearly includes rotting of non-living things. We have already seen in this section the way in which hot countries affect the growth and also the longevity of living things.[120] Here it is the living body's own heat that perishes, affecting its ability to nourish itself. It has its own heat, and engages in nutrition because of what it is. How this heat is to be understood is the subject of the next section.

If we are concentrating on what things do – in this case live a long or short time – then the factors that influence this may be and in fact are taken in a very broad way. We are interested not merely in the difference in life-span between plants and animals, and the interesting peculiarities involved, but also in the habits of animals, and humans – the randiness of sparrows, sickness or hard work – all of which lead to the loss of vital heat and wetness. More strikingly, Aristotle does not even try to define longevity, perhaps out of caution, but perhaps because he thought it was obvious what it is, in the sense of an answer to the question: how long do species, or groups within species live?[121] But within each of these groupings he clearly thought ageing a moveable feast, susceptible to many factors.

Asking how long things live is an obviously quantitative question. Yet the interesting question what one would be measuring only arises when one has begun to think about living things in quantitative terms. Aristotle does not do this. It is presumably connected to this blind spot that he nowhere makes any attempt at defining what long and short life are.

Aristotle says near the end of *LBV* that he has now given the causes of long and short life.[122] This might be taken to imply that there is no such connection between *LBV* and *JSVM* as I suggest: the enquiry is concluded with the explanation of the explananda. However, *LBV* is not, nor does it claim to be, the whole explanation of living things. Nonetheless, it fits into such an account, namely *de Anima* and *Parva Naturalia*; it would seem strange for Aristotle to have an account of longevity independent of his account of life. The next sentence refers forward to the four phenomena to be explained in *JSVM*: youth, old age, life and death. There is an obvious connection between length of life and the stages of life, in the capacity of

94

living things to preserve life, even if Aristotle himself does not make the connection here. We shall now take up the interpretation of *JSVM* where we left it at the end of the last chapter, and discuss the role of heat in nutrition.

4.4 Fire, connate heat and perishing

Aristotle has attempted to identify the central organ, guided above all by its functions of, nutrition and perception (see above §3.5), but as yet he has said little or nothing about how it performs them. This is the next step he takes in pursuing the causes of life and death: we must follow him in discussing the *symphyton thermon*, 'connate heat', which is situated primarily in the central organ and which, as we have seen, plays a central role in all mixtures, but more especially in concoction, the process whereby living body is renewed. It is through this heat that living things primarily nourish themselves.[123] Now that we have seen that living things have as their matter the capacities hot, cold, wet and dry, in the rest of this chapter we can consider the organs involved in preserving these qualities. In this way it will become clear how the body is both beneficiary and instrument of nutrition, as we have claimed.

Rather artificially we interrupted our treatment of *JSVM* after chapter 3 to interspose a lesson on Aristotelian capacities in living things, and to see how these capacities affect length of life. We broke off in the middle of the arguments for the presence of the controlling organ in the middle of the body. Chapters 2 and 3 offer a series of proofs based on phenomena (*phainomena*) for the location of the nutritive and perceptive soul.[124] We have a further series of arguments in chapter 4 based on the principle that nature does things the best way possible (469a28-b6, *kata ton logon*). Because of this, and because the centre is best for control (469a32-b1), things *must* be organised in a certain way.[125] It is the matter that is necessary and it is the form that it is necessary for. The functional parts of living beings are hypothetically necessary for the performance of those functions. Thus animals, which are defined by their possession of perception, must have this situated in their heart if they are to be well organised. For its localisation there allows for good functioning.

Clearly this distinction into arguments based on perception or phenomena, and those based on a principle is of major methodological significance. It seems to suggest that Aristotle realised that when talking about good organisation (chapter 4), one is doing something different from looking at what happens in vivisection (chapters 1, 2), or the order of generation (chapter 3). And that seems a reasonable conclusion to draw about his procedure here. But that does not turn the principle that nature does things the best way possible into a mere principle guiding reflection. This is clear from two considerations. First, the arguments from vivisection and generation in chapters 2 and 3 have some teleological basis. Secondly, principles in natural science are derived from experience of many cases by abstraction. Of course, one may say, reflection is needed for this process of abstraction; but the reflection does not involve denying that the principle determines the way things are.[126] Thus we are justified in saying both that the teleological principle guides our thinking about living things, and that it determines their structure.

The functions of taking in food and working on it will be best carried out if the nutritive and perceptive soul are in the middle. The argument from functional structure here only in fact requires that the parts that receive food (e.g. the mouth) and those that finish working it up (e.g. the stomach) will work best if they are both near to, and hence under the control of, the central organ.[127] Being close to both of these parts, the nutritive soul has control over both. In the language we met in *JSVM* 1, the central part has power over the other parts of the body, that is to say can change them. One of the further arguments in chapter 4 for a separate primary part for the nutritive soul, has it that the used and what uses it should be different, not only in their capacity, but also in their locality, as a hand is distinct from the flute it plays. In the comparison, the user is the hand, and what it uses is the flute.[128] This is then applied to the capacity (sc. of the central organ), occupying a different place from those parts of the body over which it exercises control.[129] One part is responsible for causing the change (growth, nutrition) in the other parts. In such an arrangement, the part causing growth is the dominant organ. This seems to provide for a difference between the seat of the nutritive soul and the other parts of the body, and this difference is the basis of the exercise of control by one part over another. Thus we return to the requirement stated at the start of the search for the seat of the soul: the part in question has 'capacity', i.e. power, over the rest of the body.[130] Furthermore, since the user and the used are specialised parts of the body, we may again be dealing with an argument based on the idea of superior organisation. In such cases where the principle is in the middle, and so locally separate from the parts it controls, there is organisation superior to those cases where such specialisation is not present. If that is the reasoning behind the separation between user and used, it would underline the suggestion made in our discussion of *JSVM* 1 (see above §3.5), namely that the real concern behind the idea of a unitary central organ is that of complex organisation. A central organ using and controlling the other parts of the body is the sign of highly developed living beings. Put simply, greater complexity is only possible between parts with the relevant capacities.

In *JSVM* 4 he argues from his view that food is worked on, 'concocted', by heat in the heart, to the presence there of the soul, including the perceptive soul. This heat is connected to the soul, which is, as it were, kindled in the heart. So the source of this heat must be in the heart (and analogue) since, while all parts concoct food with natural heat, the governing part does so most of all. Hence, while living things can survive the cooling of the heat in their other parts, if the heat in this part is cooled they perish entirely.[131] One of the main difficulties in talking about connate heat is its relation to the soul; and it is crucial that Aristotle says here that the soul itself does not catch fire, it is merely 'as though' it did.[132] For Aristotle is clear about the difference between soul and what is necessary for soul: the requisite heat is kindled, and not the soul. Something else is present in the heart besides the heat: the actuality the capacity serves. The soul is not heat but works through it.[133]

The second half of *JSVM* 4 supports the claim made above (§3.4) that *LBV* receives further grounding from the enquiry pursued in *JSVM*. For here we are given reasons for living bodies appearing hot when alive, the contrary when they die:[134] all the parts of animals have connate, natural heat. Heat and moisture were identified as the causes of longevity in *LBV* 5. There, no reasons were given for the bodies being wet and hot, although *de Anima* II 4, III 1 and III 12, 13, as well as parts of *de Generatione et Corruptione* II 8, had prepared us for it; as had the detailed biological treatises. In *JSVM*, the function of this heat and the way it is present in the body is described: the soul nourishes the body using heat which has to be preserved.[135]

Heat is produced in the heart insofar as food fuels the heat in the heart. But this does not mean that this heat is transmitted to all parts of the body. The heat in the heart produces nourishment, which in its final stages is blood, which is transmitted to the rest of the body:[136] 'Every part works on and concocts food with its natural heat.'[137] This nourishment keeps the heat going everywhere, like logs on a fire. So, although all other parts depend on the work of the central organ – in sanguineous animals, on the heart for blood – each part ultimately concocts its own food with its own heat.[138] Flesh, for example, possesses capacities for change, including heat, which work on the food provided to it. Remember that the central organ, e.g. the heart, has power (*dynamis*) over the other parts.[139] The other parts suffer from its action, that is, undergo the change of being nourished by it. The capacities for change in uniform parts and the organs made of them act on and undergo change through one another.

It is the heat of the body that controls and hence preserves its wetness. This is a very important argument against Gill's view of the soul's control of the passive capacities of the body. In her view, the thing that holds together an inherently unstable thing, a body which strives in different directions, is the soul.[140] The soul is an active capacity. Thus for her something hot, such as some fire in the flesh, will be a passive capacity, held in place by the soul.[141] However, it is completely mysterious how something hot can be held in place, i.e. kept from flying off upwards, by something which is neither hot nor cold. In Freudenthal's view it is the vital heat that does the work of preserving tissues. The advantage of his view is that it makes comprehensible how something hot can act on something cold, or something hot on something wet.[142] As long as the hot is there to master the wet, the composite stays together, but because Freudenthal sees vital heat as a stuff with form, he cannot explain the presence of the soul in the same account, as we find in the *Parva Naturalia* and the *de Anima*. To understand the function of the heat we must understand it in terms of the activity of nutrition: heat consumes food, and this consumption is regulated. Otherwise we are left with a mysterious form-bearing element called 'pneuma' and a superfluous soul.[143] If we think of the soul in contrast to our sand-castle, the activity of building, and the form guiding it would have to be one; the soul is activity, and regulates activity.

The final sentence of chapter 4 may be translated:

Thus life must coincide with the conservation of this heat and what we know as death must be the destruction of this heat.[144]

Clearly, this heat must be understood in the context of the nutritive and perceptive soul, which chapter 4 is concerned to locate. The heat in the central organ is necessary for life; or, in the classic formulation of the relation between matter and the end it serves: if there is to be life, there must be preservation of heat,[145] and so death is the destruction of this heat.

This is the first approach to a definition of life and death. But a crucial factor determining life and death – the need for the preservation of fire and heat in general – is not broached until chapter 5. For this reason chapter 5 is the introduction to the rest of *JSVM* and its descriptions of the mechanisms of preservation. It is the connate heat, especially that in the heart on which the heat in the rest of the body depends, that will have to be preserved. Since preserving something means preventing it from perishing, a treatment of preservation must describe the perishing prevented.

Before going on to discuss the kinds of perishing, we must establish what it is that perishes. For there is an argument among the commentators which is central to the interpretation of chapter 5: does Aristotle mean fire or connate heat, when he discusses perishing in *JSVM* 4 and 5?[146] This is relevant to a wider and more fundamental issue: how is connate heat related to other heat, and in particular fire? There is a passage relevant to this problem in *JSVM* 14. That chapter gives us a resumé of the importance of heat for living things as described in the first five chapters; our discussion may approach it using the notion of vitalism. Vitalism might be taken to hold that there is some substance, unique to living beings, that accounts for their life. In Freudenthal's interpretation, this is 'vital heat', present in pneuma. The resumé in chapter 14 begins by saying that '*some* heat' (*thermotês tis*) is needed for life, since digestion requires soul and heat is needed for soul:[147] the simplest reading is that it is not just any heat, but connate heat, the heat associated with some life: corpses cannot be revived by warming in front of the fire. They have to have their own heat. But this does not mean that this heat, *as heat*, is any different from other heat; indeed, to perform its function in living things it must behave in certain respects just like any other heat. If we can show how this works, there is no need to regard Aristotle as a vitalist; we will not have to see connate or vital heat inhering in pneuma as a special life-giving stuff, but merely as the heat of living bodies.

If we bear in mind the way he thought of fire, its closeness to the heat in a living body is more apparent: 'fire is an excess of heat and like a boiling.'[148] This is important for a variety of reasons; an excess of heat should be relative to some norm, and it seems not impossible that the norm here is the rest of the body. The heat in the heart is excessive. Further- more, as we will see, there is indeed a boiling in the heart, namely the final concoction of blood, and Aristotle speaks there of fire (see below §4.7). Without implying that connate heat is an excess of heat *because* it is fire, let us point out the similarities. *JSVM* 5 anyway provides material for thinking that connate heat is like other heat: on the face of it, Aristotle is talking of fire in general, as well as of ageing and death.[149] The latter can only be caused by the loss of connate heat – only, that is, when a living being cools down. That is to say, certain processes such as cooling are analogous in fire and connate heat. The salient feature distinguishing connate heat from other heat is that the animal has innate heat as soon as it comes to be, and then produces it itself: it does not come from outside e.g. from a fire.[150] Fire and connate heat are the same in the relevant respects, which is just as well considering the many comparisons Aristotle makes between them: both can be cooled, both rise and raise other things with them, both must be fed. Later in *JSVM* these characteristics of the parts of bodies will play a leading role. Aristotle does not ask about the constituents of fire and connate heat. That is not the subject here. In appealing to the case of fire, he is obviously trading on something more familiar to make his point about the obscurities of living beings, but that does not mean that the heat in each case is different.[151] As heat, it must be the same in each case. That was part of the lesson of §§3.4 and 4.1 above: the capacities of ingredients remain in products, even if they temper one another.

Now that we have established that Aristotle talks of fire in general *and* connate heat perishing in *JSVM* 5, let us turn to the two ways in which they perish. Aristotle calls them exhaustion (*maransis*) and extinction (*sbesis*).[152] Both kinds of perishing happen through the same cause: be- cause of a failure of food. Food or fuel is consumed so more must be provided. First, in the case of extinction, the fire perishes because the opposite, i.e. cold, prevents the fire from using the food. This could happen because of freezing, or because the living body is cut up and so loses its warmth.[153] This seems easy enough to understand. If you put something cold on a fire, it goes out; living things freeze to death in extreme cold. The other case, that of exhaustion, is harder to understand: a fire goes out if it gets too hot. Furthermore, the relevant sentence is obscure:

> When too much heat has accumulated, in this way it soon also exhausts the food, and finishes consuming it before the [process of] evaporation can be formed.[154]

The basic idea is that there is so much heat that it consumes all the fuel available and hence cannot continue burning. The process of evaporation (*anathymiasis*), seems to be the process by which the liquid food burns or boils, giving off vapour. We can perhaps understand this kind of exhaustion best by thinking of a fire that burns too fast for the fuel that is being delivered to it. An added complication is that Aristotle goes on to say that this process of fuel exhaustion explains why a large flame consumes a small one: the large one takes away the small one's fuel: the lamp's flame is simply burnt up in a large flame. Before the lesser flame's fuel can be burnt by the lesser flame, the greater flame has already consumed it. Michael offers a useful way of seeing the comparison. He suggests that if you throw a lamp into the fire, the oil is consumed and the flame goes out. Food is like the oil thrown onto the fire. The lesser flame is quenched by the greater since being a flame is to come to be continually 'like water coming to be from snow [consists] in the snow being continually melted. If there is no matter, i.e. food from which it comes to be, [the fire] is necessarily quenched, in the same way as water ceases coming to be if there is no snow from which it comes to be.'[155] If this interpretation of burning is right, it fits the view of food as the matter which is worked on by digestion. The matter must be in continual supply so that the connate heat can continue coming to be, i.e. burning. This is of course a process of transformation: the fuel becomes fire. This is the meaning of 470a4-5:

> The fire is continually coming to be, that is (*kai*), flowing like a river, but it escapes our notice through its speed.

In turn this means the fire is in continual need of food, new matter.[156] This forms the end of the passage describing the way in which a large flame consumes a small one, by taking away its fuel – a case of exhaustion. This passage is clearly a Heraclitean reminiscence, containing as it does some of the most familiar elements of his thought: flowing, rivers, fire, coming to be.[157] The difficulty is that these ideas are not usually seen in the context of the explanation of living things. But such a connection might be seen as follows. Fire exists by being constantly supplied with fuel, since otherwise it gets too hot and burns out. So the fire is like a river flowing, continuously coming to be. Similarly, living things must feed their connate heat continuously, since otherwise it burns itself out.[158] This stops them from becoming cold.[159]

Aristotle offers us a familiar example for the necessity of the preservation of heat: the contrasting cases of a brazier with a lid, which goes out quickly, and the coals banked up with ashes for the night, which stays alight for longer. The lid stops air getting to the fire and cooling it, and hence the fire with a lid is extinguished more quickly, since it consumes its matter. This is an example of extinction. Taking the lid off and putting it on brings air to the fire and keeps the fire burning by preventing the quick consumption of the matter. This process of allowing air in contrasts with 'hiding' (*krypsis*) the fire at night using ashes, so that the embers glow all night.[160] This illustration should not make us think that Aristotle thought there to be such a fire in living things: he is trying to make a point about the extinction of heat using an example more accessible than living beings.

Connate heat is being discussed because of what it does, its activity. Hot, living bodies nourish themselves. Cold, dead ones do not. The activity of nutrition explains why living things have to be hot. Thus we must, as against Freudenthal, insist on the primacy of actuality. It will be remembered that he thinks that vital heat, as a stuff, is the sole explanatory factor for vital functions. However, although the actuality of nutrition always is in matter, it is prior to the matter. The reason that connate heat, and indeed the organs that contain and preserve it, are necessary for living things is that they nourish themselves. The heat in the heart is introduced as continuing the explanation of function. Indeed, this is why the discussion in *JSVM* has to be seen as a continuation of *de Anima* II 4: only when we have established that living things must nourish themselves, and how soul and heat are connected in that process, does it make sense to explain that vital process further in terms of connate heat's liability to perish and consequent need for preservation.

Further reasons for Aristotle wishing to see an intimate connection between the principle of nutrition and that of perception are to be found by considering the constraints imposed on living things by their capacity to perceive.[161] Nutrition preserves the dynamic balance in living things, and the capacity of perception is constituted by such a balance in the relevant parts (obviously, not every balance constitutes a capacity for perception, e.g. a very earthy mixture as in plants). And it is simpler to have that part which is chiefly responsible for preserving the balance as the chief part of the perceptive system. An important passage is to be found in the discussion of the relations between the senses and the elements (*de Anima* III 1), with the aim of determining whether there may be any further senses. Aristotle's strategy is to align senses and elements. Fire is allocated either to no sense or to all, 'For nothing is capable of perception without heat'.[162] This view of perception provides an additional motivation for the desire to see the seat of perception and nutrition as one: nutrition preserves a balanced heat, and this heat plays a role in perception. The perceiving body preserves itself, insofar as it nourishes itself. There are many connections between the idea of a balance or mean between capacities and perception, some of which at least deserve mention. At the end of the *de Anima* (III 12, 13), Aristotle returns briefly to the nutritive soul in his account of the functions necessary for things to live:[163] everything must have nutritive soul as long as it lives, for food is necessary for growth, prime and decay. Perception is necessary for animals to move, he says, and this in turn is necessary for nutrition.[164] Here we have a hierarchy of faculties: they all serve the nutritive faculty, hence the introduction to the topic through that faculty. Within perception there is also a hierarchy of faculties: touch is basic to all animals and because of the kind of body that the sense of touch requires, animal bodies have to be of a particular kind.[165] This is particularly relevant to the question of the seat of the soul, since the central perceptive organ is identified with the organ of touch.[166]

The argument for the necessity of touch to animals is as follows.[167] An animal is a living body, and all body is perceptible by touch. If the animal is to survive, it must have touch; for if the animal is touched by another body, and has no sense perception, then it will not be able to seize some bodies and avoid others. And if it does not do this, it will not survive. Because sensitive beings have bodies they must have touch. For danger and good will be encountered in the form of other bodies. This provides the reason for taste being a species of touch, namely that sense of touch for those tangible bodies that are nutritive: food, a beneficial tangible body, and hence one to be pursued, is perceived by taste as a form of touch.[168] In chapter 13, Aristotle argues that the sense of touch must be in a mixed body;[169] the fact that animals must have touch has consequences for the kind of bodies they are. We have already met arguments for the presence of the capacities (hot, cold, dry, wet) arising from all four elements in mixtures (see above §4.1). The reasons there were based on the cohesion of the body concerned. Cohesion in a mixed body is only possible if the capacities in the mixed body are in balance with one another. The reasons we now find are of a different order, as we are now dealing with living bodies: now, the reasons concern the functions of living things, namely perception.

To understand how touch requires a body with certain capacities, we must understand something about Aristotle's view that perception is caused by the thing perceived acting on the animal. Each sense is only acted on in such a way as to cause a perception by its proper object – sight is acted on by visible things, touch by things that are tangible. Because any sense organ must have the capacity to be acted on by the relevant quality, the sense organ of touch must be as variable as its objects, and such that they can act on it.[170] Touch is a sense that can distinguish not merely the differences between dry and wet, but also hot and cold. Thus the organ of touch must itself be not only dry and wet, but also hot and cold, if it is going to be acted on by the proper objects of touch, which include all these qualities. Now we are in a position to understand how touch requires a mixed body. If we are to perceive whether something is hot or cold, we must be hot and cold, i.e. we must be somewhere on a range between hot and cold.[171] The idea is that we only feel something as hot or hard because our body possesses a certain hotness or hardness, and it is through the contrast between the quality of the thing felt and the quality of our body that we feel the thing as hot or hard.[172] Thus our body can be placed on a scale somewhere between hot and cold, and also between wet and dry. It is the basic qualities that belong to bodies as such, and so in any case must belong to the organ of touch insofar as that is a body.[173] The argument that touch requires a mixed body is also used to explain why some parts of animals, such as hair, and also plants, are not sensitive: they are of earth. That is, they are not qualified as somewhere between hot and cold, dry and wet. They have no mean quality which would enable them to feel.

The conclusion to Aristotle's argument here about the composition of sensitive bodies is that none of the elements can be the body of an animal: the organ of touch is not (merely) earthy. That is to say, bodies with a sense of touch have to be mixed. For touch is the mean of all tangible qualities and the organ of sense is receptive not merely to the differentiations of earth but to both hot and cold and all the other tangible qualities.[174] Only when it is clear that all perception depends as such on touch can an argument about touch establish the necessary constitution of an animal's body. The other senses require touch, and touch requires a mixed body.[175] It is the argument about touch that establishes that animals cannot be simple bodies. In the final section of the chapter further indications are produced for this vital role of touch:[176] only excessive qualities in the field of touch destroy the animal. Something very hot will kill an animal. With the other senses, all that is destroyed is the sense organ. Too much light blinds one, but does not kill. This emphasises the role of the mean in perception: if the body has the sense of touch by being *this* hot, for example, then heating the body up too much will destroy this sense. But of course, if perceiving something can destroy the balance of capacities, then perceiving must in some way involve a change in the internal balance of the animal. The point is that the perceptive parts are flexible; they can undergo a range of alterations in their ratio. They can return to their mean state because the balance restores itself; unless, that is, the change has been too great, thus over-stretching the self-restoring balance. It is for this reason that Aristotle is so interested in those cases of perception where the capacity is destroyed by the perceptible: the mean is so affected that it is destroyed. But how does the destruction of touch destroy the animal? Once the organ of touch is destroyed, there is no more sensation, and hence there is no more animal, since all animals have to have sensation.

In the context of *JSVM*, the obvious point concerning the central organ is that balance must be preserved there if there is to be life, and since this balance is preserved by nutrition, nutrition must continue. That is to say, to see how touch is preserved, we have to ask why living things stay hot. 'Hot' means some quality in the range between hot and cold. That is to say, 'hot' is to be seen as a mean. Since touch requires a mean, and a mean is to be understood as a balanced state between extremes on a quality range such as hot-cold, the process of nutrition, in preserving the balance of the living thing, also preserves the capacity for perception. Precisely the same balance or ratio that nutrition preserves constitutes the capacity for perception. You can regard this actual nutrition as the capacity for perception. Thus he is able to say that the principles of perception and nutrition are the same in number, merely different in their being. The conception of balance plays a central role in two ways. First, all complexes are constituted by such a ratio. And secondly, the burning of the psychic fire (as a special case of the first use of the idea of a ratio) requires that it be neither too great nor too little. When the balance is lost, either through violence or the natural effects of the environment on its constitution, the living thing dies.

The capacity to grow can be understood in comparison with, and in contrast to, a fire. The capacity to grow lays hold of things, like a fire, and changes them.[177] But unlike in a fire, the heat in the homoiomerous part is tempered by the admixture of the other elements. Thus the heat in such a body does not convert the matter supplied into fire, as a fire does, but into the kind of mixture that it is in. The heat has an assimilatory-discriminatory power,[178] working on the (passive) food. Sweet and potable matter in food is drawn by innate heat into flesh and other parts of the body, while bitter and salty are excluded.[179] Heat divides the similar things off, making them like, and leaving the unlike behind as residue: it takes those capacities in food that are suitable for (e.g.) flesh and makes them into actual flesh.

Connate heat digests. The word Aristotle uses for 'digest', *peptein*, *pessein*, means 'cook, bake'.[180] In the context of Aristotle's biology it is often translated 'concoct'. Aristotle's definition of concoction runs as follows:

> Concoction is maturity, produced from the opposite, passive characteristics by a thing's own natural heat, these passive characteristics being the matter proper to the particular thing. For when a thing has been concocted it has become fully mature. And the maturing process is initiated by the thing's

own heat, even though external aids may contribute to it: as, for instance, baths and the like may aid digestion, but it is initiated by the body's own heat. In some cases the end of the process is the thing's own nature, in the sense of its form and essence, as when moisture takes on a certain quality and quantity when cooked or boiled or ripened[181] or otherwise heated; for then it is useful for something and we say it has been concocted.[182]

If we begin the interpretation of concoction from the end of this definition, we can see that in both cooking and digestion the relevant point is that the product is useful. In the case of natural things, presumably primarily living things, their own heat matures the matter, i.e. food, by working on it until it has turned into a product that is part of that thing's nature, e.g. the right kind of flesh.[183] For the idea of something hot having a peculiar function such as digestion, we can compare the heat of an oven. It is only heat, but it also cooks. In order to do that it has to be nourished (fed), but also tempered so that it does not burn the stuff to be cooked, and does not go out too quickly by using all its fuel; one way of expressing this might be to say that its heat is balanced. The oven has to be organised in such a way as to ensure these things. In order to approach the idea of concocting food as an activity, consider the oven again. What is it doing? One answer is: it heats things. But giving this answer shows that you do not understand what an oven *does*, what its function is. You use ovens to cook things. The heat needed in an oven is relative to cooking. Similarly with the heat e.g. in the heart. It is not only hot, it also serves as the means for the animal to digest and perceive. The hot matter, i.e. parts, of a living being are relative to, and hence posterior to, the functions they perform.

In chapters 4 and 5, we have a theory of the heat in living things which is made necessary by its function. One crucial fact about this heat which is relevant to the way it functions in living things is that it perishes. It does this in two ways: either it burns itself out by being too hot, or else it gets too cold. This suggests that we should see these two ways of perishing in terms of the heat having to be neither too hot nor too cold, if it is to survive.[184] That is to say, a mean heat, a balanced heat, must be preserved. The heat must be preserved from getting too hot by being cooled, and from getting too cold by being nourished. This means that the difference between survival and death is that between balance and imbalance (within some perhaps broad range). To understand this balance we have to bear in mind the function of the heat which has to be so preserved. The heat has to be preserved in such a way that it can nourish the whole body: the account of perishing must refer to that of life, that is to say, to the vital processes. The need for preservation of the innate heat provides the motivation for the rest of *JSVM*, which describes first the preservation of heat in plants and then the much more elaborate account of that in sanguineous animals through breathing. That is to say, the cooling mechanisms investigated in the rest of the treatise should all be considered as ways of preserving the balance within the whole organism. They preserve a balance, this time not in the tissues, but in the whole organism.[185] There is, moreover, a further level of balance which will have to be taken into account, namely that between the living things and their surroundings.

JSVM 7-27, i.e. what is usually called the *de Respiratione*, is purely about the cooling necessary for the preservation of heat and hence life.[186] That is to say, these chapters are dependent on *JSVM* 1-5. In this way, *JSVM* can be seen as a single treatise describing how living things, and in particular sanguineous animals, preserve the balanced heat in their bodies, and so their capacity to nourish themselves.

4.5 Balancing heat

Now that we have seen, quite generally, where the soul is seated, and how it is related to heat, living bodies must be looked at to see how they prevent this heat from perishing. When that has been established generally, we will be in a position to generalise about living things. So this enquiry requires looking at the broadest categories of living beings in Aristotle's view. *JSVM* 6-19 describes the way in which heat is preserved in balance, using fairly detailed physiology. Such preservation is a common activity of all living things, but there are a variety of ways in which this function can be performed: by the environment, by food, by breathing and by water passing through the gills. Thus we are dealing with the 'common and peculiar activities of living things', as the introduction to the *Parva Naturalia* suggests we will be (see above §3.1). *JSVM* 5 showed the necessity for heat to be preserved from perishing, in particular from exhaustion from too much heat, by cooling; respiration is only one of these ways of cooling things, the most sophisticated and that nearest to us. Furthermore, as has become clear, the need for cooling is intimately connected to the need for food. The connate heat must be preserved so that it can work up the food that sustains the heat of the whole organism. Thus we are basically dealing with feeding.

We begin with the most simple case, that of plants, in chapter 6. Their cooling is effected by their food and their environment. This immediately raises a problem: why does food cool down the heat, when its function is actually to keep the heat going? To explain this, Aristotle uses a comparison with what he thinks happens when humans eat: when food is first ingested and before digestion happens, it causes the air in the body to move. Apparently, the turbulence in the air that food causes in the body is meant to work like a cooling breeze, and so is comparable to respiration.[187] Only when the food is digested does it contribute to the heat. Until the logs freshly put on a fire actually catch fire, they cool the fire.[188] Aristotle also compares what he thinks happens when people fast: they grow hot and thirsty. That is to say, they will cool down if you give them food. Yet there is a difficulty here, in that we would tend to offer someone a glass of water, and not a piece of dry bread, if they were hot and thirsty. But if we look more closely at Aristotle's idea of nutrition, we will find that food and moisture are very closely related. For him, nutrition includes moisture. This can be seen from his treatment of taste. Taste is the sense related to food, and the thing tasted must have moisture in it. For tastes are in moisture as their matter. So although we may think of food as solid (as opposed to water), Aristotle obviously did not; and we may think of the obvious cases of blood and milk. Taste is always the taste of food and you only taste something moist. This connection is also underlined by the fact that he nowhere mentions the need for sufficient moisture for living things. This makes it clear that he included the wet in nutrition.[189] In this way we can understand how Aristotle thought that food cools things down.

Because they are so dependent on their environment, plants are subject to seasonal withering, or drying out.[190] They wither in the cold because innate heat is cooled by the cold, and in hot weather because food is burned up.[191] This is thus parallel to the two ways in which heat can be destroyed (exhaustion and extinction) described in chapter 5. Both can be understood as forms of imbalance, but this time between the environment and the plant: the climate is too hot or too cold for it.[192] It is noticeable that Aristotle devotes almost all of his attention to the case of the loss of balance, rather than to the proper functioning of the plant. Perhaps this is because the supposed cause is more evident when things go wrong. A plant being dried out in high summer is more obviously a case of excessive heat than its flourishing is a case of balance.

Plants, unlike animals, use the heat from their surroundings to work up their food. In Aristotle's picturesque phrase, they use the earth like a stomach.[193] There are, however, further problems with the explanation of growth and withering by the change in the seasons.[194] It may work for annual plants to say that they grow with the heat (as long as it is not too great) and wither with the cold. But how, for example, can something perennial continue to grow throughout several years? The answer might be that it has not lost all its heat at the end of the cold period: there is still enough of it which has heat to grow and nourish itself. We can compare this chapter with *LBV* 6 and its elaborate account of how plants renew themselves. In contrast, *JSVM* 6 is basically negative: such is the concinnity of plants with their surroundings that they need no particular arrangement to keep them cool. This makes clear that the effect of the seasons can be understood merely as a contributory cause to the growth and decay of the plant; primarily, we must consider the nature of the plant concerned. Merely by being living bodies that nourish themselves in their environment, they are kept cool. The extreme simplicity of their arrangements is indicated by the story Aristotle tells to explain the fact that pieces of broken pottery are put at the roots of plants by some peoples. This, being earthen, that is to say, cold and dry, keeps the roots cool in hot climates and so preserves them from withering.[195]

The final sentence of the chapter points to the two habitats of animals, air and water, which are used to effect cooling in respiration. This is the further project of the *JSVM*, to deal with both water and land animals, as far as the cooling, and hence the preservation of their heat, is concerned. This point draws attention to the importance of the environment for the preservation of the balance in living things. In many cases, but not all, Aristotle thinks that this happens through cold stuff (air or water) drawn in from outside to cool the internal heat. Plants are dependent on their environment for their cooling; but so are animals, even if this dependence involves action on the part of the animal. What this action is, and how the animal body can perform it, is the subject of the rest of *JSVM*.

As one would expect from the subject of the *Parva Naturalia*, Aristotle's interest in breathing here lies in its function, rather than merely in the relevant organs:[196] we do not understand breathing until we know what it achieves. This question is the starting point for his criticism of his predecessors: some did not ask the question; those that did, did not answer it correctly. Unlike nearly all his other treatments of his predecessors, this one (*JSVM* 7-13) is presented as an interlude. The motivation given is not to construct his own position, but to avoid the charge that he is accusing them emptily.[197] However, one might think that this is slightly disingenuous of Aristotle: his debt to his predecessors is clearly massive. Thus this piece of doxography does have a close relationship with other, and more famous, examples in the *Corpus*.[198] The following are features which appear in his critique of his predecessors, which are central to his own treatment of breathing:

the function of breathing (7 470b7, 11 472b24);
the connection to life and death in Democritus' account (10 472a10 f, 11 472b27-9);
the connection of breathing to causing life heat (11 472b32, 12 473a3, 9 ff).

Even this superficial glimpse shows how closely Aristotle's own theory is related to his historical context. The reasons given for the predecessors' mistakes are also significant: lack of acquaintance with the inner organs and no acceptance of final causes.[199] These two aspects are clearly related in Aristotle's thought, in that knowledge of the constitution of the organs and their interrelations makes sense for him in terms of a project investigating what they are for.[200] The perceptual qualities of organs, e.g. the sponginess of lungs, betray their capacities in the functioning system. Aristotle also emphasises that breathing is not a general feature of all animals. He argues against Democritus' view that all animals breathe; insects are cited as a counter-example.[201] Breathing thus cannot explain all death since not all animals breathe. This explains the caution expressed at the start of *JSVM*: life and death occur to only *some* animals through breathing and its cessation.[202] The 'peculiar and common actions' of living beings do not refer to a list of different actions, but to the things that all living things do, although they perform them in peculiar fashions. For example, all animals need cooling, but only some of them achieve this by breathing.

The end of the excursus is marked at the end of *JSVM* 13 474a23-4 by saying that he has given the *difficulties* associated with their accounts of breathing. The same perspective that guided these criticisms is also in the foreground in the discussion of the different kinds of animals that breathe. Despite a bias in favour of large animals with blood, the idea behind the treatment of breathing is that there is a variety of ways to fulfil the need for cooling – what mechanisms they use, and what complications there are in understanding how they breathe (*JSVM* 15-19). The enquiry is general, about all kinds of living things. While we are interested in the features of living things' bodies, it is not these bodies themselves that are the object of the enquiry, but the actions they can perform: cooling the connate heat. This must be cooled to preserve the activity of the nutritive soul. This not only picks up the work done on locating the central organ in chapters 1-4, but also concerns the further explanation of longevity: things that are good at preserving their heat live longer.[203] None of this was discussed in *LBV*, but is an obvious continuation of the explanation of longevity.

Let us now look at some examples from chapters 15 to 18 to illustrate how his consideration of living things illuminates the way they function. Plants have already been dealt with, however cursorily, in chapter 6. Chapter 15 starts off the account of cooling by approaching it with the most general groupings of animals that Aristotle uses – land and water animals, sanguineous and non-sanguineous animals. The use of these classifications is to ensure the generality of the account.[204] The immediate relevance of these groupings lies in the fact that cooling is related to the environment of the animal,[205] and the possession of blood is a mark of heat. Chapter 15 is about animals with little or no blood, i.e. those that are not very hot. He begins from the simplest animals – ones that are most similar to plants in that their heat is not very great, and so is sufficiently cooled by the surroundings.

Aristotle is interested in comparing longevity in the different kinds of animals: life is prolonged by efficient refrigeration.[206] Since this would seem to explain things said in *LBV* 5, these observations are very germane to the question whether *LBV* can be fitted into the complex account of living bodies in *JSVM*. There we were told that life is hot and wet, and death is cold and dry; we were given the functional ground of this difference: heat is needed for nutrition. Here in *JSVM* the physiology of cooling, and hence the preservation of the hot and wet, is explained. So *LBV* is not isolated: further grounding to the account of longevity is being given through physiology. It must be admitted that these connections are not explicit, but the role of the hot and the wet in the functioning of living things is so fundamental that making the link explicit is necessary for us, not for Aristotle.

Aristotle views living things as arranged in an order of 'honour'; but there is no direct connection between 'honour', i.e. complexity of organisation, and length of life: more complex animals do not simply live longer. It is, however, clear that soul explains heat: living beings are hot because they must feed, and nutrition requires heat. Thus Aristotle is not in fact in the embarrassing position that Freudenthal's reductionist approach would put him in.[207] If Aristotle had tried to explain length of life purely in terms of vital heat then there should be some correlation between the amount of vital heat and length of life. Yet because he is able to take into account the differing structures of animals and plants and the functioning that these structures make possible, he can be happy to have long-lived primitive beings, such as trees. There are many embarrassing aspects to vitalism, and one of them is that it seems to make the structure of living things irrelevant; Aristotle's lively pioneering interest in organs shows how far removed he is from this view.

There is, however, some correlation between the complexity of the living thing and longevity. The shortness of the life of bloodless animals is due to the fact that their heat is nicely balanced: you do not need to do much to it to make it go out. Aristotle expresses this by saying it has no great momentum (*rhopê*), i.e. downward tendency, but the word is also used generally for the capacity to move.[208] The basic image is of scales nicely balanced: only a small casting weight is needed to effect the change between life and death. In the case of cool animals, not much cooling is needed for them to die. This has both a negative and a positive side. They can be easily destroyed, but they can be easily preserved: 'only a little helpful thing is needed to save them just as only a little unhelpful thing is needed to destroy them'.[209] That is, they do not need so much heat to keep them going, but this heat is vulnerable: they possess no mechanism to preserve it. So we find again the idea of balance expressing the conditions needed for survival. This picture of a balance between two inclinations for change is obviously connected to the idea that mixtures in a living thing are in a stable balance. Stability is seen as a relation between capacities for change. The idea of living things having a momentum, greater or smaller, for living, which has to be in balance for them to continue alive, is the expression of the same idea on a higher level, the functioning of organs.

The longer-lived insects are warmer than others and do not breathe:

instead, they cool themselves using the air that is inside them (innate breath) (474b31-475a20). We have already met Aristotle's decided opinion that not all animals breathe; innate breath is, at least in part, a term invented in order to cover those animals that do not breathe, but nonetheless have breath inside them. The noise of this breath cooling the inside of the insect is the buzzing that 'warm' insects such as bees make. We see here Aristotle concerned to connect the theory of what animals have to do (cool their heat to preserve it) with what everyone knows that they do (bees buzz, cicadas sing). The same point will emerge with breathing. Michael of Ephesus suggests that the reason insects were thought not to breathe is that they could (apparently) survive under water.[210] This would explain why the movement of an insect's abdomen was seen not as evidence of breathing, but of internal cooling by internal air. We should think of insects as having a closed cooling system, like that in a water-cooled engine: no material exchange takes place between the cooling system and the environment. In contrast, animals that breathe have a cooling system that is open to the air, rather like an air-cooled engine. This point is relevant to the way Aristotle understands breathing itself: it is like a cooling breeze, which is not consumed when it cools you on a hot day.[211] It has been remarked that if you replaced Aristotle's 'need of cooling' with 'need of oxygen', you would get a theory which approximates more nearly to the truth.[212] The big difference, however, is that air is not consumed in his theory: the coldness of the air is merely destroyed by the heating of the blood. But there is no reason to think this consumes or destroys the air itself. The reason that Aristotle thought that breathing cooled is the common experience that we breathe more often in hot weather; and he thought that the cold has the effect of slowing our breathing.[213] He also noticed heat going in and out of the mouth, i.e. that we lose heat when breathing; this is even more noticeable when we pant.[214] His theory of breathing fits the reason that animals such as dogs pant, to cool themselves.

Chapter 16 moves on to sanguineous animals and introduces the two organs Aristotle sees as cooling systems: gills in water and lungs on land. We are not offered a physiology of lung and gill. On the physiology of the lung – shape, position and structure – we should turn back to *PA* III 6. The relation between *PA* and *PN* is complicated by the fact that the structure of the organs is also to be explained by their function, and hence related to soul.[215] Thus the same function of the lung is, naturally, mentioned in *PA* as in *PN*: cooling the body. There is, however, a difference between discussing the organs and other parts of animals to distinguish and describe them, even if such a discussion requires mention of the function of the parts, and discussing the functions of the parts because you are talking about the activities of the whole animal. The latter is the project in *PN*.[216]

In *JSVM* 16, Aristotle proceeds by asking which animals have lungs and then gives a fairly elaborate catalogue of animals; his interest in a general treatment comes to the fore once more. He starts with internally viviparous animals: if animals have heart and blood, then they breathe to refrigerate (475b16-19). The same applies to oviparous animals. Those with wings are birds, those with scales, snakes etc. Viviparous ones are bloodier, whereas the oviparous are drier and hence have longer intervals between breathing.[217] The final category is that of breathing animals which pass their lives in water (476a1-15). They have gills. It is remarkable that Aristotle lighted on the functional equivalence of gills and lungs. This must be because he had noticed no animal with both lungs and gills (476a6-15). One means of cooling is enough for any animal. Since nature does nothing in vain, some have gills and others have lungs, but none has both.[218]

Chapter 17 looks at another piece of equipment, vital for both feeding and cooling: the mouth. It is explained how one organ is capable of fulfilling two functions.[219] One might think that 476a16-18 says that the mouth has the one function – preservation of being – which it serves in two ways by admitting both food and air, both of which are necessary for preservation. But the rest of the chapter shows that there are *two* functions, and how this extreme of economy is possible. Feeding and breathing at the same time is not possible for breathers (476a28). At 476a5, animals with gills are described. They have no problem with the mouth since they expel water (and take it in) through the gills and only eat through their mouths. This account shows Aristotle fitting organs to functions.

Chapter 18 again shows Aristotle working in generality with his account of the cetaceans and of the function of their blowholes: to expel water taken in while eating rather than to being connected with breathing, i.e. cooling. The whole motivation for this study is provided by the project of the *PN*. The detailed physiology here is entirely relevant to the question of how, in interesting cases, cooling is effected. He is clearing misconceptions about blowholes out of the way, so fitting difficult cases into his general account. Dolphins choke in fishing nets, so they breathe air. They thus do not contravene the observation that no animal has both lungs and gills (476a14-15). The function of the blowhole is to prevent the entrance of water into the air passages more efficiently than the epiglottis can. These animals get their food in water, and to do this they must be able to discharge the water they take in with the food. The blowhole is not an organ of breathing since it connects with none of the organs with blood in them. He thinks this parallel to crustaceans which take in water to feed and not to breathe (476b30-477a10). Of course, he was wrong about the anatomy and function of the blowhole; but right in the observation, if such it was, that some cetaceans take in water with their food.

Chapter 18 refers back to Aristotle's biological treatises for details. At

477a7 he refers to *HA* (VIII 2 589b2, cf. 533a30) on the discharge of water in a variety of animals and to *PA* (II 2 648b2) on which animals are cold and which hot. On polyps, mentioned at 477a5, we can turn to the general account at *HA* IV 1-3. These references confirm that the main emphasis in the *PN* is not on physiology, but on how vital functions are exercised in various kinds of animal. We get the details elsewhere, or just need reminding of them, if we are thinking of his works on physics as a continuous course of enquiry. The vital function we are concerned with is of course nutrition, and the preservation of the heat necessary for nutrition. One has the distinct impression that Aristotle was very impressed by his own ability to explain some difficult cases, such as cetaceans, and the double use of the mouth for breathing and eating, and that these tests for a bravura explanation rather divert from the account of the actions of animals. But the display is impressive for the ordering of a mass of detail within a powerful explanatory framework. The general account of nutrition, and the cooled heat performing it, is fitted to specific cases, thus showing that such activity exists, by showing how it comes about.

4.6 Explaining breathing

In *JSVM* 19 Aristotle tackles the remaining question, how cooling takes place in animals with lungs, by dividing it into two aspects, the first relating the lung to the soul and the heat of living things, and so to their degree of development and to their surroundings, the second concentrating on the bodily structure involved, starting in chapter 21. Unlike many of his predecessors, he restricts breathing to animals with lungs,[220] and thinks that he was the first to see that the lungs had a function – hence his insistence that only animals with lungs breathe. He is thus able to assign them a function within the living thing.

In chapter 19, both material (heat) and formal (soul) explanations are used to account for what he calls the 'more honourable' animals: those with lungs, above all – humans:[221] for he says that it is necessary that animals have a more honourable soul 'along with' (*hama*) more heat.[222] We have investigated the relation between heat and nutritive soul above; it is no surprise that there is a correlation between heat and function.[223] The important thing to see about this passage is that while there is a necessary correlation between the two factors, there is no necessary dependence such that the heat, for example, would make the soul superfluous: both causal factors are needed. On the basis of the correlation between heat and (honourable) soul, he can then explain the correlation between the possession of a large hot lung and size, presumably because they are hotter and so able to consume more food, and a second relation between the unique uprightness of humans (held up by heat) and their great quantity of the

113

'purest' blood.[224] What he is doing here is integrating the lung in the explanation of what a thing is; as he puts it:

> So that the reason for this part as for any other must be posited as belonging to the substance [of the living thing]; for [the living thing] possesses it for the sake of *this*.[225]

So far teleology is triumphant, guiding the enquiry into the body, both in relation to its surroundings, as here, but also in terms of the qualities of the lung itself (chapter 21). But in chapter 19 (477a25-31) Aristotle goes on to discuss 'the cause from necessity and of change', which one should think, he says, constitutes such animals with lungs just as it does other things. Some living things like plants come to be from more earth, water animals from more water; that is, they have a more earthy or watery composition. So too with animals that are very hot. Which cause or causes are meant here? Ross[226] cites the last lines of *de Generatione Animalium* for the expressions 'necessary cause' and 'the cause of motion': there we are concerned with accidental non-functional features caused by matter, and not by the final cause.[227] Thus for example we are concerned with the colour of eyes which is not explained by the end of seeing, but by the peculiarities of the matter performing that end.[228] But, as Michael of Ephesus remarks,[229] the necessary cause for Aristotle is usually matter: in our passage we are dealing with matter serving ends. But then we may ask: is matter here both moving cause and necessary cause? This would fit very nicely with the close connection between matter and motion or change in general (see above, esp. §3.4). The problem with this reading is twofold. One difficulty can be discounted fairly simply, namely that moving and material causes are officially different. But as our discussion so far has made clear, matter is explanatory in living things insofar as it possesses a capacity for change, and the subject under discussion here *is* matter in living things as a cause of change.[230] The other difficulty is one that faces any reading of this passage: matter would seem to occur twice, first as the heat correlated to the soul (477a14), and secondly as the necessary cause here. But it is simplest to think that the heat in the first passage is being derived from its source in the hot surroundings of the animal in the second. In effect, what the second passage is doing is explicating the necessary correlation in the first: the necessary cause refers to that which is necessary for the end. This reading would align the present passage with those in which matter is necessary, if ends are to be achieved. Or as Aristotle also puts it in an important passage elsewhere in *Parva Naturalia*, when describing why sleep is necessary to all animals:

> I mean necessary on a hypothesis, because, *if* an animal is to exist, having

its own nature, then certain things must of necessity belong to it, and if these things belong to it, then certain other things must of necessity belong to it.[231]

This necessity, conditional on the existing form of the living thing, is commonly known as 'hypothetical necessity':[232] if a wall is to exist, then you need bricks (for example), and they must be of such and such a kind. This relation between the form (soul or nature) of something and what is needed for this form does not of course preclude the kind of necessity that holds when parts of animals with their capacities for change interact: the capacities for change – for warming and being warmed – interact, and the change is necessary.[233] As we shall see, such capacities for change and their interaction play a central role in explaining breathing, involving as it does the cooling of heat.

Saying that honour and heat are correlated, as Michael of Ephesus does, does not tackle the problem of the priority of the soul to the matter it is in. Indeed, he tends towards a purely material explanation: some animals are more honourable because of heat. In the case of humans the blood around the heart is lightest, purest and hottest, and is the reason for man's uprightness.[234] These views leave Aristotle claiming that the function things are capable of can be explained by their matter. Yet it is clear from this passage (esp. 477a14-25) that we have two complementary levels of explanation:[235] the function of things can be explained by both form and matter, and, as one would expect, the matter serves the form. This leaves the form free to be prior to matter. That is part of what it means to say that in Aristotelian physics material explanation is used, even in the case of the activities of living things. But, of course, the form is not eliminable. This passage would thus seem to belong with those in which a certain feature is explained as both good and necessary: in *GA* II 1, for example, he says that the division into sexes can be ascribed both to matter, i.e. a necessary and moving cause, and to a final cause, namely that they are for the sake of generation.[236] Here the presence of the lung is being explained both by heat and by the soul of the living things concerned.

Living things are surrounded by capacities for change, and must be so constituted that they are not eliminated straightaway; rather, they must be related to their surroundings in such a way that they are sustained.[237] Aristotle's idea is that things are made of stuff from their surroundings, thus explaining the earthiness of plants, the airiness of birds, and the heat of upright animals, and this heat lies in their possession of blood-filled lungs, allowing a balance, internal and external, through breathing.

The food that is consumed in this way must be suitable to be assimilated by the thing in question. An obvious way of viewing this suitability is to say that the matter ingested must have the capacity to be changed by the living thing (see above §3.3). But does considering matter in the context of the living things that it is in, preclude the possibility of seeing the connection with matter when it is not in living beings? It would be disastrous if it did: we must avoid the danger that matter becomes miraculously modified on becoming part of living things. On the one hand, it has to make sense that *this kind* of matter should go towards forming part of the living thing, because of the capacities of this matter. On the other hand, matter must be modified when it comes to be in a living thing. Something happens to hay when the horse eats it. But such matter (food) is only one aspect of what we may call the external balance of power of the living thing. For if the food is part of these surroundings, then it will in some sense fit in these surroundings, just as the living things in question must. Neither may be worked on to an extent that destroys them; conversely, ingesting food is only an aspect of the support offered to the living thing by its surroundings. It is presumably also affected by the heat and cold around it.

Part of the explanation why some animals have lungs involves discussion of the necessary cause and the cause of motion, which make such animals as man, and others, come to be: some living things such as plants come to be from more earth, water animals from more water.[238] So too with animals that are very hot. This is the first indication that we are concerned with the surroundings here: water animals, i.e. animals at home in a watery environment, come to be out of more water. Yet it is not a completely new subject since this cause is clearly related to the material cause, i.e. the heat which provided the reason for the lung. Aristotle is concerned with the lung as explaining the posture of animals in relation to their surroundings: things that are upright must have light upper parts above the heavier things underneath them. Heat and more especially the heat in their lungs explains this relative position: heat rises. This answer, however, implies a view of the way living things are related to their surroundings, and this explains why, as part of the explanation why living things have lungs, Aristotle embarks on an elaborate criticism of Empedocles in chapter 20. There, he argues, against Empedocles, that things live in suitable environments. That is connected to the point that water animals come to be from more water.

Empedocles' theory is presented in terms of excess and deficiency: living things are opposed to their surroundings so that they can balance their excess or deficiency against the corresponding deficiency or excess of the environment: for example, Aristotle claims that Empedocles thought that fish had migrated from land to water to cool their excessive heat.[239] Instead of such a simple opposition between surroundings and living beings, Aristotle thinks there must be a congruence in the material, which, however, does not exclude an opposition in their states, that is, in the modified matter in the living thing (478a6 ff). On this view, Empedocles is wrong because contraries destroy one another; if there were hot fishes in water, they would soon perish. The connection with the preservation of the animals is evident: if Empedocles were right, they could not even begin to preserve themselves. The relation to the environment forms the basis on which the physiological equipment of animals can be effective. Aristotle compares having something made of wax in a hot place (477b18). Instead of there being a contrariety of *matter* between the living thing and its surroundings, there can be a contrariety of *states and dispositions* of the matter.

It is very important here that the matter is identified, as it has been in *LBV* 5, as the hot, wet, dry, cold.[240] It is on these terms that a living being can have a nature which is saved by being in its proper place, by being wet in water for example.[241] The idea seems to be that a fish comes to be from more water. The capacities of the water – wet and cold – are preserved (but not the water), and predominate in the fish: fish are mainly wet and cold, and so can survive in a wet and cold environment. Yet because such qualities as wet and cold can change to being more or less, for example something wet can become wetter, the living thing can become too wet for its own good: in relation to its nature it becomes too wet. It is this state of the fish that can be in opposition with the surroundings: the surrounding water might cool the fish down too much. The surroundings – water, air, earth and fire – will act on the living body insofar as their heat, wet, cold and dryness are in tension with these capacities in the living thing. So if the surroundings are too wet for your constitution, you might become sick.

Not surprisingly this account fits very well with the idea of balance, which was the way the criticism of Empedocles was introduced. There has to be a balance not only between the capacities in the living body, but also between the living body and its surroundings, if the living being is to survive. Furthermore, there can be interaction between these two balances: if the surroundings are very cold, this can affect the balance within the living body. Since one pair of states of the living body that can be caused by balance and imbalance is health and sickness, it is also fitting that we can see in this passage (478a2-7) a reminder of this topic which emerges sporadically throughout the work: bad states of an animal can be improved by a change in the season or place. This idea of correcting unbalanced states is best taken as a reference to health and sickness: a change of surroundings brings us back to balance, i.e. health.[242] When the state of the body is excessively hot, a cold place will restore the balance. Obviously, going to a cold place will not mean that the nature of the living thing changes from being predominantly hot, as it is in man, to being predominantly cold.

An obvious way of understanding 'states and dispositions' of matter here is by using the idea of mixtures: as developed in §4.1 above this would mean that the capacities from the ambient elements are modified by being tempered by other capacities, while still ensuring a basic harmony between living thing and its element. This conception strains the idea of a

117

mixture to the extreme, in that mixtures seemed to require some kind of mean between the ingredients. Yet the present idea requires that some element preponderate, so that fish, for example, survive in water but not on land. But the point presumably passes for want of any awkward measurements of proportions; or indeed, of any awareness of the possibility of such a measurement.

Aristotle sums up his view as follows:

> With regard, however, to states of body, a cold situation has ... a beneficial effect on excess of heat and a warm environment on excess of cold for the region reduces to a mean the excess in bodily condition. This must be sought in the regions appropriate to each type of matter, and according to the seasons which are common to all; for while states of the body can be opposed in character to the environment, the material of which it is composed can never be so.[243]

Thus we see that the preservation of balance is the key to survival: if fishes were hot, they would never be in balance with their surroundings and so perish. In order to offset an imbalance with its surroundings, the living being must repair itself with food etc.[244] There are two ways in which this discussion of the constitution of living bodies and their environment is connected with *LBV*. First we have here the idea that living bodies are acted on from outside. In *LBV* 3 (see above §4.3), the contraries which cause perishing were either in the body or outside it. This connection with the environment goes back even further in *LBV* to chapter 1 (465a8), where Aristotle touched on the effect of hot places on longevity. Here we have something of an *explanation* of the relation between the two. Although living beings and their environment are not radically opposed to one another (i.e. in their matter) there may be an imbalance between them. For the living thing can become too hot or too cold in respect of its environment. This account relates finite things to their environments in terms of their capacities for change, thus leaving their activity unaffected by their environment.

How does this discussion of the relations between living bodies and their surroundings explain the presence of lungs in some animals? This discussion arose from consideration of the necessary cause and the cause of motion (19 477a25) which are to be taken as the matter from which living things come to be. Since the lung is hot, the conclusion must be that things with lungs come to be out of hot stuff. The heat is preserved as a capacity in such things and is as such the material cause of the lung. Since such hot living beings are in hot surroundings, there can be a balance between them and their surroundings. After this relatively abstract discussion of the balance between animal body and its environment, Aristotle turns in chapter 21 to the question of the constitution of the body itself:

the body is not merely hot and wet, it consists of organs which themselves have a complex structure. Since, however, the organs in question are concerned with the interaction between environment and living being, the nature of the environment remains relevant. Chapter 21 begins the more narrowly physiological part of the treatment: what must the lung and air be like to be able to breathe and to be breathed respectively? The main points made in this chapter concern the sponginess of the lung (478a13), its very bloody character and the lightness of air, which is thus able to penetrate the lung (478a18-21). Furthermore, the lungs are full of blood: it is obviously the blood that has to be cooled. That is what air does in the economy of things with lungs, and its lightness is the reason that it can perform this function.

In the previous chapter, 20, we had an extensive discussion of the way living things have to be in balance with their surroundings, if those surroundings are not to destroy them. But this balance can be preserved by the internal organs of the living thing: this is the added sophistication of sanguineous animals. We have here, as already in chapter 15 (474b30 ff), the device of the small *rhopê* of vital heat used to explain a feature of living things: their heat can be preserved or destroyed by a small influence. This was used to explain the short lives of small bloodless creatures, since they have no way of their own to preserve their natural heat (see above §4.5). In the case of sanguineous animals too, their heat is said to have a small momentum (*rhopê*).[245] This is because their great heat is vulnerable to exhaustion. The treatment of the mechanism of breathing continues in chapter 22 and is dependent on the reason that refrigeration is necessary, namely the 'kindling' of the soul.[246] It is connected to previous chapters in *JSVM*: to chapters 5 and 14 because of the necessity of cooling for the heart, and also to chapter 19: we are looking for the material cause of breathing. But it also goes all the way back to the first four chapters on the place of the central organ (cf. 478a34-b1), where the soul is 'as it were kindled'. We already know that the lung is there to cool the heart (chapter 21), but we still need to know what the connection between heart and lung is, at least the way changes in the one can affect the other. There must be some link between the two of them if the one is to cool the other. At 478a27-8 Aristotle refers to *HA*, but also to dissections concerned with the connection of lungs and gills to the heart:[247] the anatomy is not the subject of *JSVM*, although it obviously is to be assumed. Ogle makes the following comment:

> Aristotle does not mean that there are open passages by which air can pass directly from lung to heart but merely that there are blood vessels connecting the heart with the lung and allowing the air to pass through their walls from the contiguous air tubes.[248]

Below we shall consider whether air does in Aristotle's view actually pass into the heart in breathing. The reason this question is important is that we have to decide how Aristotle understands the relation between inhaled breath and other 'connate' breath in the body. The alternative to inhaled breath being consumed is that air cools simply by passing in and out of the lung. This is the importance of the fact, noticed in chapter 21, that the lungs are full of blood: blood and air come close to one another in such a way that the blood's heat heats the air. While Ogle makes clear that there is no open passage for air between lung and heart, he does assume that there must be some way for the air to pass into the blood. Aristotle does not actually say this: all he speaks of is water passing through the gills, and animals with lungs drawing in and exhaling breath (478b13-15). The same applies to water and gills: the water just passes through the gills, and we are certainly not to think of water cooling the animals by passing into the blood. So on the basis of this passage, there is no reason to think that air enters the body, i.e. the blood vessels, via the lungs in respiration. It penetrates no further than the spongy lungs, and is then exhaled.

Further evidence for this interpretation may be found in Aristotle's reaction to a debate about why one breathes: some people had suggested that breath is a form of nutrition for the fire within (473a3 ff). Aristotle attacks this view in *JSVM* 12,[249] reasoning that there would have to be some analogue in other animals that do not breathe, since all animals have vital heat, and that we see that such heat arises from nutrition. And the last reason he gives against breath being food, is that then food and waste product would use the same channel, and he denies that we see this happening elsewhere. That Aristotle is wrong here should not blind us to the fact of just how wrong he is. He is denying in effect that breath is consumed when it is breathed in. To repeat, its function is purely to cool by its passage. That is why innate breath in bees and suchlike can cool purely by its passage.

The end of chapter 22 is best seen as a summary in preparation for the main phenomena considered in the rest of the treatise, all of which are centred on the action of breathing. The account of the chest moving up and down in breathing, and of water moving through the gills (478b10-15), reappears in chapters 26 and 27. Suffocation, caused by a lack of air and hence cooling (478b16-19), is dealt with in chapter 25;[250] both old age and disease make breathing impossible, and so cause death, as is explained at length in chapters twenty-three and twenty-four (478b19 f). The main topic of chapter 27, respiration, is fundamentally concerned with the preservation of balance, namely between hot and cold. Respiration happens because the hot matter in which the nutritive principle is seated increases, and hence rises.[251] In fact, this perpetual generation of heat is part of the activity of the nutritive soul, since the increase in heat comes from food, which thus preserves the nutritive heat.[252] This heat then lifts the breast cage, expanding the lungs, which thus are forced to draw in (relatively) cold air from outside. This penetrates the spongy tissue of the lung (cf. chapter 21) to the vicinity of the hot blood, cooling the chest which then sinks, expelling the now warmer air. Aristotle offers a comparison of a familiar artefact with the less familiar explanandum: namely the bellows.[253] This comparison is vital insofar as it makes clear the connection between expansion and drawing in air, contraction and expelling air. Both lungs and bellows expand when drawing in air, and contract in expelling it. In the case of breathing, this double motion causes a rising and sinking of the chest, in Aristotle's view. He obviously has a very particular kind of bellows in mind, since another point of comparison between lungs and bellows is meant to lie simply in the shape. Quite how the bellows in question looked is unclear in the absence of illustrations; but they were 'double' in some sense. In the case of the lungs, this shape helps them perform their function, because they can enclose the heart and so cool it from both sides.[254]

The connection between the account of breathing and life is quite explicit:

> The inward passage of air is called respiration, the outward expiration, and this double movement goes on continuously as long as the animal lives, and keeps this part in continuous motion; it is for this reason that life depends on (*esti en*) the passage of breath outwards and inwards.[255]

For animals that breathe, existence is not so much like breathing as dependent on it. Conversely, it is clear that the function, nutrition, that breathing is implicated in, is also continuous (see below §4.7). This is not the definition of life: that is given in chapter 24 as the remaining of the nutritive soul (see below Chapter 5); the passage merely defines respiration so that it is clear how life depends on it; and hence how respiration must be treated in a work on life and death. Aristotle also deals with vital cooling in fish: they have no lungs, so the account of cooling must be completed by discussing gills (480b12-20). The movement in gills is like that in the lungs. They too rise because of the heat and let water through, they contract and expel water. That this is a rather less felicitous account than that of lungs need hardly surprise us, nor indeed that he basically allows himself to be guided by what he thought he had discovered in the case of sanguineous animals. Just as life depends on breathing for land animals, so for fish, it depends on admitting water.

Before moving on to the treatment of the heart, let us quickly recapitulate this and the last section on *JSVM* 15-22: these chapters describe how cooling is carried out in animals, that class being divided into subordinate kinds. In some cases, this is effected, as in plants, purely by environment and food. The shortness of the lives of such animals is due to the sensitivity of their vital heat to outside influence. In others, the more developed ones – sanguineous animals – there are distinct organs (lungs and gills) with the function of cooling in the main divisions of animals – land animals and water animals. The vital importance of the lung in cooling and so preserving the living being leads Aristotle first to consider the causes of having lungs: these are the amount of connate heat in the animal and the function that it exercises. This leads on to the question of the heat of some kinds of animals as compared to others – water animals as against land animals. While living beings are in tension with the outside world, they cannot be composed of a matter that would itself be in opposition to their surroundings. For this would make the task of self-preservation impossible. Furthermore, these organs are distinguished from other, related organs (blowholes), and one of the ancillary organs (mouth) is described. How it is that the lung is suited by its tissues to being the organ for breathing, and how it is connected to the heart, and so can cool the blood there, have also been described. The physiology described in these chapters serves to explain the action of breathing and nutrition. In *JSVM*, we have progressed from the abstract consideration of the parts of living things (top, middle, bottom; front, middle, back) in order to locate the central organ, to consideration of the mechanisms necessary to cool the central heat. This is a characteristic move in an Aristotelian enquiry, from something true and general, to an articulated understanding of the varieties falling under the generality. These mechanisms preserve the balance both in the body and between the body and its environment. Yet neither have we seen the central organ in action nor have we been given completely general accounts of the subjects of *JSVM*; together these occupy the rest of the treatise.

4.7 The heart and pneumatosis

In this chapter we have followed the capacities in the body from their simplest level in the uniform parts to the organs involved in preserving the natural or vital heat of living things. In the last section we saw how the lung provides the cooling necessary in the most developed beings, so that their heat does not consume itself through excess. Aristotle's approach to the process producing heat is through suffocation and then the motions he associates with the region of the heart (479b17): palpitation, pulsation and respiration. Breath, both inhaled and internal to the animals, plays an important role in these discussions. We will try to show that it is not necessary to interpret Aristotle as constructing his account of living things using some vitalist notion of 'pneuma'. Pneuma is instead a stuff arising like any other from the interaction of stuffs with capacities.

Suffocation is a striking phenomenon connected with breathing and the distinction between animals with lungs and those with gills: those with

122

4. Body

lungs choke in water and the others choke in air. Aristotle tries to explain this phenomenon in chapter 25, as occurring when cooling is forcibly stopped through a lack of the right coolant, water for fish, air for animals with lungs. The reason that the coolants do not work lies in their nature, and in those of the lungs and gills: the wet cannot penetrate through the whole lung quickly enough, while water animals are not quickly enough cooled by air because of the dryness of gills.[256] The organs of refrigeration are suited to the environment from which the refrigeration is to be drawn. The refrigerant must be suitable for there to be refrigeration. Like the extensive treatment of death in chapter 23 (see below Chapter 5) and the use of vivisection in chapters 1 and 2 (see above §3.5), the attention paid to suffocation shows Aristotle deriving his account of life from a view of what is happening when things break down. Not only is the connection between life and breathing manifest, given what happens when someone is drowning, but it is also clear that the peculiar nature of water or air is connected to the preservation of life through breathing or the function of gills.

Chapter 25 also provides some difficulties for understanding the arrangement of the last chapters. At 479b13-16 Aristotle promises an account of how lungs and gills move, drawing in and expelling water or air. But instead of actually explaining how lungs and gills move, he gives an account of palpitations, pulsation and breathing.[257] However, these motions are a necessary part of the account of breathing: until we understand what the distinctions are between these things, we cannot understand the motion of the lungs. For these motions have been thought to be the same (479b17-18). The three 'things that happen' (*symbainonta*) around the heart are: 'palpitation', 'pulsation', 'respiration' (*pêdêsis, sphygmos, anapnoê*). The descriptions of palpitation and pulsation are not just preliminary to that of respiration: they add to our knowledge of the workings of the body and soul. They are concerned with three areas close to the preoccupations of the *Parva Naturalia*: health, sensation and digestion.

Palpitation involves cooling and occurs for one of two reasons: either because of disease or from fear. Both of these causes make it very interesting indeed: the way disease as a cause of palpitation is introduced here suggests that normal functioning is health. Fear is interesting because it is a sensation: and most of the time in the *LBV* and *JSVM* we have been dealing with animals without much being said about their sensations or perceptions, although these are characteristic for this kind of living being. The importance of palpitation for understanding the *Parva Naturalia* has been noticed by Theodore Tracy in his remarks on a famous explanation of anger at the start of *de Anima*:[258] the material explanation of anger is a boiling of blood around the heart. Tracy remarks that this links the heart's function as thermal centre of the organism with its function of emotional response and movement. For it suggests that if the bodily aspect of anger

123

is an abnormal increase of vital heat in the heart and its blood, the bodily aspect of the opposite, fear, must be a cooling down of the blood of that organ. This ties in precisely with the detailed physiology offered in *JSVM* 26.[259]

The second kind of movement around the heart is pulsation,[260] which is vital to the account of living and dying since it concerns nutrition; but it presents considerable problems because of its close connections with breath. What happens to liquid food under the influence of heat is *pneumatôsis*,[261] and it is this that causes pulsation. Aristotle compares the whole process with the throbbing of an abscess, which continues until the concoction is complete; this throbbing is painful because unnatural, morbid.[262] In turn this concoction is compared to boiling liquid in a vessel, which expands and escapes as steam. This last aspect is clearly meant to illustrate how the concocted food, i.e. blood, escapes from the heart along the blood vessels. The boiling is perhaps more than just a comparison; as we have seen,[263] fire is 'an excess of heat and like a boiling'; the liquid concocted here is of course blood.

As the name suggests, pneumatosis must be connected in some way with pneuma, i.e. breath. If the account of breathing given above is correct, then no breath enters the heart; it penetrates the lungs but no further. What then is the breath in pneumatosis? An obvious answer would be connate breath (*symphyton pneuma*). This is, put roughly, breath that is in living bodies, and is not introduced into them by breathing. How to take this innate breath is, however, disputed: as a vital force,[264] or as merely one of the material factors in living beings. If we are to understand pulsation, we must make a short excursus to deal with innate breath.[265]

In order to show how innate breath is in fact best taken as a material factor, it is useful to distinguish it from innate heat, which has become familiar to us through its role in the preservation of life (see above §§3.3, 4.4). This problem distinguishing innate breath and innate heat arises because in one text it seems that breath is necessary 'for internal heat'.[266] But this is a misunderstanding: breath is necessary for internal heat, and is not to be identified with internal heat. Breath does not sustain the internal heat by being hot itself; it preserves heat from perishing through an excess of heat, i.e. from exhaustion.[267] So this text refers to the cooling of internal heat by breath.[268] This function of cooling heat must be taken on by innate breath in those animals that have no lungs and therefore do not breathe.[269] Finally, there are the two best known functions of innate breath: it serves as a tool for the formation of the heart and other organs, and it provides force for mobile animals.[270]

If the above account of pneumatosis is correct, then innate or connate breath is hot because it arises from heating liquid. As to the functions of internal heat, we have already seen how the heat of living things is involved in nutrition (see above §4.4). This function is connected with various other duties this heat performs: it makes living things hot as long as they are alive, and has its principle in the heart or its analogue in non-sanguineous animals. Its presence in all parts of the body enables them to digest the food that reaches them.[271] It is also involved in sensation, thought and movement, and causes sleep by carrying bodily matter, i.e. food, to the organ of primary sense and so making it incapable of functioning. The mechanism of sleep deserves a slightly more extensive description because of the closeness to the treatment in *JSVM*: the cycle of waking and sleeping is determined by the process of nutrition. Blood and its analogue in non-sanguineous animals are the final nourishment.

> When food enters the receptive places from outside, evaporation happens into the veins and there, changing, is turned into blood and proceeds to the principle.[272]

That is to say, food enters through the mouth into the stomach, is converted either there or in the vessels attached to the stomach into blood, and then proceeds to the heart, where it has an effect on the animal's capacity to perceive, since the perceptive soul also has its principle there.[273] Aristotle wants to know what kind of incapacity sleep is, as opposed to, for example, unconsciousness. Sleep occurs because of the evaporation (*anathymiasis*) concerned with (*peri*) nourishment. Roughly, what happens is that the vapour rises, is cooled by the brain and sinks back to the middle of the body, stupefying the common sense organ with its cold.[274] It takes some time for the nutritive soul, using the heat in the centre, to digest that food. The comparison with a fire is interesting in view of the problematic relation between fire and the heat of living things:

> When much food is introduced, which heat drives up, like fire when logs are put on, [the heat] is cooled until the food is digested.[275]

This text confirms the similarity between common or garden fire and the natural heat of living things suggested above (§4.4). When the food has been digested – separated into its different parts by heat – the animal awakens. This description of sleep touches a central topic: the relation between nutritive and perceptive soul[276] – and it should be clear that at least part of this connection lies in how hot the central organ is. In all these texts, the connection is material: we sleep not because of something that connects the function of perceiving with that of feeding, but because the matter we ingest stupefies our perceptive organ.

The view that innate breath is a vital force is relevant not only to the question of the survival of living things, but also to the question of whether this survival is connected to the balance of capacities within the body. For on this view, innate breath preserves the living body by establishing a balance between the other constitutive factors of the body. As Nussbaum puts it, innate breath 'guarantees the body's organic unity' and preserves the 'specific equilibrium'.[277] The key to this line of interpretation is most easily understood if we think of Aristotle answering the question: how do bodies composed of four elements stick together? While earth and fire move in different directions, they are held together by nutritive soul; and it is innate pneuma, acting as a tool for the soul, that performs this task.[278]

This line of thought runs against the whole interpretation of the composition of living bodies and the activity of the nutritive soul that has been developed in this book. The action of the heat in living bodies is easily seen as nutrition, and this preserves the equilibrium of the hot living body. This balance has its roots in the relation between the simplest capacities in the uniform parts, although no actual elements need be mentioned in such an account. There are, however, further objections specifically concerned with the loss of balance, and with the question of how such a loss causes destruction: living beings do not perish because their bits fly off in different directions. Rather, they die because of functional failure, which has to be explained by Aristotle on the level of the organs concerned and fitted into the life-cycle; it is this task he is tackling in *JSVM*, which has been largely overlooked. While the idea of balance, which pneuma is meant to secure, is central to Aristotle's conception of the preservation of living things, this balance is not secured by a mere stuff, but by organs in action.

This short excursus should have clarified the following terms:

(1) Breath (*pneuma*) is breathed in and out, does not enter the body beyond the lung, and cools the heat in heart and lung (see above §4.6).

(2) Connate heat is heat in and arising in the living body (see above §§3.5, 4.4).

(3) Connate breath (or pneuma) is breath that is in the body and arises in the body. In animals with blood, Aristotle describes this process in detail as the boiling of blood in the heart.

After this treatment of innate breath, we are in a position to turn back to *JSVM* 26 and its treatment of pulsation. In Aristotle's view pulsation occurs when the half-worked-on food from the stomach comes into contact with the heat of the heart and is thus worked up into blood. Thus he closes his account of pulsation by saying that it is 'the pneumatosis of the wet being heated'.[279] During this process the food is pneumatised; we note that there is no mention of breath as it is breathed in playing a part in this process of pneumatosis. So if it is not the inhaled breath that is involved in pneumatosis, what is? It is internal to the animal in that it is not introduced from outside. A very simple interpretation of the process of pneumatosis is all that is necessary: when a liquid in a container is placed on sufficient heat steamy vapour is produced. This is part of the process of cooking the liquid – Aristotle himself makes the comparison with boiling. When the food hits the heat of the heart it increases in bulk from dense to rare, like boiling water being pneumatised (*pneumatizesthai*) by the fire heating it.[280]

The pneumatosis of blood enables it to cause nourishment and growth by giving the blood *motion*.[281] When the raw food hits the heat in the heart, it goes off as moist vapour (*anathymiasis*) and feeds the other parts of the body.[282] This moist vapour is the blood boiling over, and so escaping from the heart along the blood vessels to the other parts of the body.[283] The process in the heart is evaporation (*anathymiasis*). This account enables us to understand pneumatosis as a necessary part of the process of working on nutrition. A further aspect might lie in the separation of differing parts of blood. Although admittedly this goes beyond the text, it is attractive to think that part of blood is pneumatised and thereby separated from the coarser stuff, which is left as it were as sediment, while the hot vapour rises to the head.[284] But why is there *beating* of the heart? Think of the noise and commotion that a vigorously boiling liquid produces. In a viscous liquid, this may take the form of a regular thumping in the pot. It is this thumping that Aristotle thought the heart communicated to the blood vessels. While we call the rhythmical throbbing of the arteries the pulse, Aristotle uses the same name for it as for the heartbeat, 'pulsation', suggesting that he missed the fact that the two do not coincide in their timing. Beating connects the heart to feeding since it is caused by the liquid food being finally worked up into blood in the heart. The *beating* itself has no function.[285] It is comparable to our way of thinking of heartbeat: beating is not the function of the heart. Rather, pulsation for Aristotle is a concomitant to a necessary action of the heat, namely the final concoction of nourishment to feed the rest of the body.

How are the lungs and the heart connected here? If the action of the lung is to cool the heart, there clearly has to be some connection between them that allows this cooling to happen.[286] One way would be that the cool air is actually introduced into the heart, but in fact Aristotle only speaks of the air touching the heat in the lung (480b4), and of each (air-filled) tube in the lung having a blood vessel alongside it.[287] This would seem to be a good arrangement for cooling blood without mixing it with air. Rather than air entering the heart, blood enters the lung. In the process of feeding no breath is consumed: it merely cools the region of the heart, and is not mixed with the liquid food at all. After all, the breath comes out again. This cooling process is more like being cooled by a passing breeze than by drinking a glass of cold water. The boiling of liquid food on touching the heart is not to be confused with breathing: that is the lesson of chapter 26. And there is no reason to think that Aristotle thought of inhaled breath as an ingredient in the concoction of blood (just as there is no reason to think that he thought that water enters fishes' blood through their gills). Rather, it is simpler if boiling is something that happens when heat gets into wet stuff, and is an intermediary state of matter between wet and fire, that is, where the capacities hot and wet are in balance.

Aristotle describes the heart beating and the lungs breathing as follows:

> In the heart, the increase in bulk (*onkôsis*) arising from wet food through heat produces beating, when it rises against the outermost mantle (*chitôn*) of the heart; and this always happens continuously, for the wet flows in continuously, from which blood's [complete] nature comes into being (it is primarily manufactured in the heart) ... The nutritive part of the natural capacity must be in the middle. Therefore [this part] rises when it becomes more, and when it rises it is necessary that the part surrounding it also rises – which clearly happens in breathing things; for they raise their breast through the principle, which is in [the chest area] of the [surrounding] part, doing just that. For when [the chest] rises, it is necessary for it to draw in air from outside, which, through being cold, cools the excess of the fire. Just as, when the fire increases in size, this part rises, so when it decreases, it is necessary for this part to sink, and when it sinks for the air which had entered to leave again (it enters cold and leaves hot through contact with the heat present in this part, above all in animals with a bloody lung).[288]

Here Aristotle is explaining what the heart does in two senses: it beats, and the reason it beats lies in what it really does: it provides a vessel for the continuous boiling of food into blood. This boiling is a fire, 'an excess of heat and a kind of boiling', and performs a central job of work towards the actuality or end of nutrition; the soul is not fire.[289] This in turn is linked to the cooling function of the heart, in that as raw blood arrives to fuel the burning, the fire increases, so raising the breast, and so expanding the lungs which then draw in cool air. This cools the excessive fire and the breast can sink; and so on continuously. The necessity mentioned here several times is that of the interaction between the parts, understood as capacities for change – expanding, contracting, cooling, heating, raising, sinking. What Aristotle does not say, but probably means, is that these changes are necessary if the activity of nutrition continues, and if it is to continue.[290] The continuousness of breathing and heartbeat, and their role in actual nutrition, are important indications of the correctness of our thesis that Aristotle takes nutrition to go on throughout life.

The diagram describes the continuous cycle of pneumatosis and breathing. Taken as a whole, it represents a central portion of the activity of nutrition taking place; the activity appears, of course, nowhere as an element in the diagram. The changes making up the stages in the diagram are coterminous with life, and support the end, the soul, insofar as, when taken together as they are in the living body, the soul is the end they achieve, the activity they perform. And it is this very end that is perpetuated. Looked at from another perspective, the cycle of concoction and

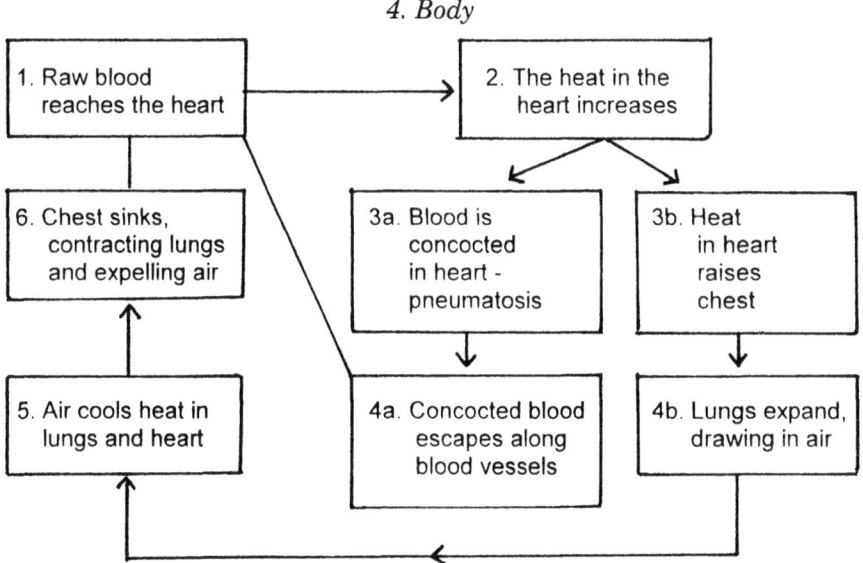

The cycle of pneumatosis and breathing in *JSVM* 26-7. Arrows mark an efficient causal connection. The lines without arrows between steps 6 and 1, and 4a and 1, indicate that the earlier step must occur before the later one can take place, but without the earlier step acting as an efficient cause for the later. Death occurs because the lung cannot perform steps 3b, 4b, 5 and 6, so causing the failure of the heat responsible for steps 1, 2, 3a and 4a.

breathing perpetuates itself as long as the state of the organs for breathing allows this. It is constituted by changes caused by organs possessing capacities for change and is thus an example of nutrition; a central feature of this self-preservation is the balance preserved between hot and cold.[291] Breathing and pneumatosis form the culmination of *Parva Naturalia* in that they are the nutritive process of large warm-blooded animals, which serve so often in the work as the model, implicit or explicit, for all living things.

5

Defining the life-cycle

Throughout this book, we have been concerned with Aristotle's explanations of life using matter and form. And we have seen how he introduces his investigation into living things by defining them – to talk about certain things, we must first delimit them. And in the ensuing enquiry we have been through the classes of living things – broadly, plants and animals, with the latter divided into those with blood and those without. But the stages of life apply to all these. When we understand something, we know its causes; in other words, we can define it.[1] Causal definitions are one kind of definition for Aristotle,[2] and ending an enquiry is one function of one kind of definition. Towards the end of *JSVM* (chapters 23 and 24),[3] Aristotle gives causal definitions of coming to be, life, youth, old age, prime of life and death. These are of great interest not only because they represent the conclusion of Aristotle's enquiry, but also for their purely methodological aspect; however, I shall concentrate on the former.

The first of these two chapters prepares the ground for the actual definitions; it starts by saying that coming to be (*genesis*) and death are common to all things, but it is the varieties of perishing[4] that the chapter deals with – violent and natural perishing, depending on whether the origin of perishing lies inside or outside the living thing in question.[5] Because his aim is to give definitions of the stages in the natural life-cycle, one important preliminary is to fit in the single part of the life-cycle that can obviously happen unnaturally.[6] Natural death is part of the life-cycle, but perishing can happen in other ways as well. In immature things, such as eggs and plant seeds, death (i.e. perishing) takes place in a different way. What is common to all perishing is that it happens through a loss of heat;[7] this allows us to have a general account of all perishing in living things, i.e. a definition.

In mature things, the loss of heat happens in the part in which the 'principle of substance (*ousia*)' is situated.[8] Obviously, if such a part has not formed, as in a seed which has not begun to grow, heat cannot be lost there. That is the part Aristotle took such care to identify in *JSVM* 1-4 (referred to here, 478b34), situated in the middle between up and down. That was, as we have seen above (§3.5), a general treatment, showing where the principle of the soul is situated in all living things. This treatment gave rise to the first provisional definitions merely of life and death:

> So it is necessary that living and the preservation of this heat are present together; and that so-called death is the perishing of this [heat].[9]

These are only provisional insofar as the conservation of heat had not been explained in chapter 4 – the forms that extinction takes, and the consequent necessity of breathing are introduced in chapter 5 and explained generally in chapters 6-22. Without cooling the other parts of the life-cycle cannot be defined, as we shall see. These observations go some way to justifying the 'therefore' with which chapter 23 opens: the definitions offered flow from the enquiry, even if only loosely. In chapter 23, two steps make up the description of death or wasting away. The first concerns the heat of living things (478b33-479a7), the second the cooling of this heat which is necessary to its preservation (479a7-28). Loss of heat is naturally caused by the lack of cooling.

The distinction between immature and mature living beings is important because heat can be lost even when the central part in which the principle is situated has not yet grown; nonetheless such living things die through loss of heat, and, of course, only die violently. Mature living beings all perish in an analogous fashion: through loss of heat in the central part. Ironically, this fact about natural perishing is established by vivisection, and in fact a similar use of such evidence was used to locate the central organ in chapter 2.[10] Which part is vital to life is discovered by seeing which parts continue living when cut off, a point which distinguishes plants and animals, although some of the latter, like plants, can survive when cut up (479a1-7). This short tour of the broadest classes of Aristotle's living world – plants and animals, the latter then divided into blooded and bloodless – shows that generality is a central concern here. In chapter 5, we learnt that there are two kinds of destruction of connate heat, exhaustion and extinction. The extent to which all perishing is the same is that it is always explicable by loss (*ekleipsis*) of heat. Violent death occurs when the lung is affected from outside, so that it cannot function any more; natural death is only different in that the cause of this affection does not come from outside the living thing, but from the organism itself.[11] Violence includes more than just the effects of, for example, knives; for Aristotle, anything which goes against the nature of the being concerned is violent,[12] and so includes sickness.[13]

In order for life to be preserved in highly developed beings, the heat in the central organ has to be preserved by cooling; and when this heat is no longer cooled, it exhausts itself, so causing natural death.[14] The failure of cooling happens through the hardening of the lungs or gills, as the case may be. Thus the immediate cause of death is not so much the extinguish-ability of heat – that it consumes itself – but the failure of heat's preservation. The reason for the failure of heat's cooling here is the hardening and 'growing earthy' of the lungs or gills which prevents their motion and hence their performing the function of cooling.[15] This probably means that the texture of the lungs and gills becomes less flexible, and so unable to function. Finally, some increase in the heat happens, and the fire exhausts itself.[16]

By this, Aristotle explains the fact that in old age not much is needed for someone to die. If it is the loss of heat in the lungs or gills that renders them earthy, then their loss of heat could be explained by 'most heat having been breathed out through the length of life' (479a16-17). The idea is apparently that breathing simply removes heat from the living body: exhaled breath is hot. This lessening of the connate heat in the heart makes even less of an increase (*epitasis*) in the heat necessary to bring about death.

In the next chapter (479b1), we learn that violent death can also happen through exhaustion (*maransis*) as well as through extinction (*sbesis*). This should be taken as a reference to suffocation treated below in chapter 25: when things are forcibly deprived of cooling, their heat exhausts itself.[17] Natural exhaustion happens when there is no such outside interference, and the heat exhausts itself. Another detail added there is that instead of merely saying that such change happens over time, Aristotle says that exhaustion is natural when it happens 'through the length and fullness of time'. This fullness, or perhaps better maturity, can be seen as the state of something as prescribed in the life-cycle after such and such changes: through its natural development. As we saw in Chapter 1 (above §1.3), the idea of a life-cycle is the reflection of the existence of a form in time: it has to be reached by growth, and is then lost.

Since the extinction of such a small light needs no 'violent disease', the old die painlessly; the parting (*apolysis*) of the soul is imperceptible.[18] So it seems that there is a double explanation for death. First, the heat grows smaller and smaller through being exhaled. We have seen that Aristotle thought that breathing cooled because our breath is hot,[19] and we breathe more often when we are heated; the idea is that we lose connate heat, the heat we naturally produ⸌e, by this means. Secondly, the heat, such as it is, is preserved by the lung's cooling; but this preservation fails because of the deterioration in the lungs.

The crucial question with natural death is: does the passage of time alone cause it? This would be seem to be suggested by Aristotle's saying that the lungs and gills are dried out through the length of time.[20] Why do gills and lungs grow more earthy? We want a *reason*. The problem becomes particularly clear if we take Michael of Ephesus' account of natural death: this occurs when the constitution of the lungs or gills is dried up by age and can no longer perform its function, and he understands the dryness that comes from age to be naturally caused. This is natural death, according to Michael.[21] The weakness of this account is that it does not seem to explain anything. It merely says that the lungs naturally dry out in time ('dried up by age'), hence causing the death of the whole organism. Why must the lung dry out?

Only the hardening *of the lung* is cited as cause for death; this must, however, be seen in the context of the end it fulfils. The explanation of natural death is not merely a matter of the composition of material bodies, as is the general phenomenon of perishing. The elaborate arrangement of living bodies and the ends they fulfil must play a central role. We need to have done quite a lot of Aristotelian physiology to get us this far. This alone might make us suspicious of a sudden return to a more basic level of explanation than the parts of living bodies; quite apart from the fact that Aristotle does not mention such a reductionist explanation when giving his causal definitions of the life-cycle; and there is no explanation for why the lower level of elements makes itself felt at some stage rather than another.

Further reasons against this view are to be found in our argument[22] that it is the capacities in the bodies that cause them to behave as they do: it is the capacities of the lung that are in question here. The capacities do not belong to elements, but to the living body. And indeed, only so can we speak of natural death, in the sense of one proper to the nature of the living thing. Otherwise, the nature referred to would be the nature of the elements, since it is their tendency to change that is responsible for the change. On our account, however, we can say that lungs and gills have such capacities that when they are worked on, after a while, they become hard. 'After a while' can thus be made innocuous: some changes require a certain amount of time to happen. Hardening of the lungs *by old age* must refer to the things that living things do; in this sense, only the living age. So there would be no advantage in trying to give a reductionist account of natural death.

Aristotle's causal theory requires us to ask the question what acts on the lungs and gills to harden them. For on the basis of his theory that any capacity to be warmed requires something to do the warming, or more generally any capacity for undergoing a change must be acted on by a capacity to produce a change, there must be some active agent to make the lungs harden.[23] There seem to be two possibilities, neither of which is mentioned by Aristotle, although both seem available to him. One cause on offer would seem to be morbid (478b24-8): diseases which make the lung hard through residues (*perittômata*), or growths (*phymata*), or an excess of morbid heat which causes rapid breathing, since the lung cannot expand and contract far. But this passage should not be invoked as a description of natural death, but of violent death. Since this chapter starts by distinguishing these two kinds of death, we expect some account of

133

violent death. As described in the present passage it would occur by a hardening of the lung by affection that is contrary to the nature of the thing concerned. The effect is the same as the hardening that happens through age: 'at last when they can no longer move the lung, they breathe their last and die'.[24]

So if these lines do not describe natural death, we are still short of a cause. The simplest way to approach the problem would be to take our lead from *LBV* 3: all material things perish, either through an internal or an external cause. Internal causes mentioned there were residues, which need not be morbid; in Aristotle's view residues can be useful (such as semen), or useless (such as phlegm); in neither case are they signs of morbidity.[25] And it is conceivable that they would, as having the right amount of heat and coldness, work on the lungs in time to harden them. This heat or cold, however, could also come from outside the body, that is from the environment, the second source of change mentioned in *LBV* 3. This would be a reference to the changing heat and cold of the seasons. That the source of change would be outside the living being does not speak against the perishing so caused being dependent on the nature of the living thing. How the heat and cold affected the lungs would depend on the constitution of the lungs. The maximum functioning of the living thing would be eroded by the interaction of capacities for change. So Aristotle could have solved this difficulty had he tried. In this way the general character of having a life-cycle is inscribed into the specific functioning body we are concerned with; so failure occurs in the fullness of time.

Chapter 23 has disposed of perishing, and the complication caused by its possible unnatural (premature) occurrence. But it has done much more than that: the connection between life and heat has been used to arrive at the definition. This connection played a central role in *JSVM* 1-4, 5, and already in *de Anima* II 4 it formed the conclusion of the treatment of nutrition: everything is concocted by heat, hence every living thing has heat (416b28-9). Here, once it has been given, it and its connection with nutrition give the clue to the other definitions:

> Therefore coming to be is the first participation in nutritive soul, [which takes place] in hot stuff; and life is the remaining of this participation. Youth is the growth of the primary cooling part, old age is its wasting away, and prime lies in between the two. The end, i.e. perishing, is, when violent, through the extinction and exhaustion of the hot stuff – it may perish through either of these causes –, and, when corresponding to nature, the exhaustion of this, happening through the length and completeness of time.[26]

He reaches his definitions of life and its stages by way of the description of death.[27] And it is easy to see why he starts with the description of death: certain very noticeable things happen then, such as the cessation of breathing; we will come to another one later. Coming to be (*genesis*) here is the first beginning of nutrition and hence growth: it marks the coming into existence of a living thing that actually nourishes itself.[28] Aristotle defines this coming to be as the 'first participation in nutritive soul, which takes place in heat'.[29] This event is the first meeting of matter and form: when matter begins to nourish itself, which it does because of the soul in its heat (acting in each species in a specific way), a new living thing has come to be. Here we have the first actuality of an organic body: soul (see above §3.2).

The formulation 'participation in nutritive soul' might make one think that Aristotle is being Platonic.[30] However, if this is taken in the weak sense of explaining why an individual participates in soul in the natural course of being generated by its progenitors, rather than implying that a living thing comes to be by participating in an abstract form, then the definition has a precise point. As we have repeatedly noted, living bodies are in Aristotle's view suited to the actuality of their life. Nutritive soul can occur in warm stuff because warm stuff is suited to performing this function. And this participation would be causally explicable in terms of the life-cycles of the living things concerned: the progenitors bring it about that nutrition and so growth takes place in the warm stuff. This is more concrete, and plausible, than saying merely that life is a combination of body and soul.

The relation between heat and nutrition is at the heart of this book. These definitions show that Aristotle's physics accommodates both matter and form, that is, the matter of the living thing and its activity. What about the stages in life? The relevant text reads:

> Youth is the growth of the primary cooling part, old age is its wasting away, and prime[31] lies in between the two.

The change in the cooling part would be responsible for the change in heat of the living thing, and this, in turn, for its youth, prime and old age. This allows for an alteration in the way in which the form or activity is in the matter, and so for the presence of a cycle in the living thing. We should not, however, be deceived into thinking that such a part can be isolated as a causal factor from the end it contributes to (see above §4.6). Obviously, soul is to be read into these definitions of the stages of life, although it is absent at a verbal level.

For a definition to be useful, the terms it uses must be either defined or, in some sense, primitive. This condition points up a serious problem with the definitions we are dealing with: they use soul – more exactly, nutritive soul. The close connection between life and soul might encourage one to think that the *definienda* here fall not under the things defined by reference to something else, but among those not so defined; and one might

135

think that life and soul have some claim to be primitive terms.[32] But it is important to notice that life and soul *tout court* are not the subject here[33] – only life *qua* generated, and nutritive soul; and, although youth, old age, life and death are the subjects of our treatise – and so primary, or primitive in some sense – they are analysable insofar as they are causally explicable. Life is indeed final or primary, but finite life is certainly neither, and so is explicable in terms of its causes.

Nutritive soul was defined, at least insofar as it was causally explained, in *de Anima* II 4 (see above §3.3). It also played a role in the most general account of soul in *de Anima* II 1; for there the preliminary definition of life was that it is nutrition through itself, growth and wasting away.[34] At first blush, this might seem to vitiate our definitions entirely; for if we define life here through nutritive soul, we should not have already defined soul by using life, the latter understood as nutrition. Compare

> Soul is the first activity of a natural body possessing life in capacity (412a27).

and:

> Life is the remaining of participation[35] in nutritive soul taking place in hot stuff.

The two definitions are intimately connected. Two related resemblances may be mentioned: both are restricted to certain aspects of life; and both include reference to the body. As for the first one, Aristotle restricted consideration in *JSVM* 1 to only those capacities or parts of soul relevant to why animals are such – nutritive and perceptive soul.[36] This suggests the second point; there is no need to specify that soul is in a body because that is the kind of soul we are talking about, given that we are talking about animals and plants. That is to say, we are dealing with bodily (organic) life. In our chapter life is approached genetically: from the point of its inception. In this way, only life can be considered that does come into being, and this coming into being is a matter of the relevant 'tools' or 'organs' first in the progenitors, and then in the new living thing. This is, of course, a continuation of the approach to the soul in *de Anima* II 1: the question asked is, When is the soul present? and the answer is the same for all living things.

The second book of *de Anima* begins its inquiry into the soul with a grasp of the soul, of a partial kind. From what Aristotle says there, it is clear that we can expect no further general account of soul; but we do get definitions of nutrition (II 4; see above §3.3), perception (II 12 424a17 f) and imagination (III 3 429a1 f). In *JSVM* we are almost exclusively concerned with the first of these,[37] building on the quite general account in *de Anima*, namely that nutritive soul, defined using the actuality of nutrition, feeds the body with food using hot and cold as tools. This definition is then pursued through first locating the tool in the body (see above §3.5), and the requirements for keeping this tool in order, heat requiring cooling (see above §§4.4, 4.5). The final definition of life will not then try to go beyond the definition we had started with, but to give the general cause, not deduced from any more general statements, e.g. about warm things or about fire, but by considering the different kinds of living things, and stating what is common to all of them.

Definitions are completely general, and serve as principles. These are very strong requirements and may seem unrealisable. But there are clear indications that Aristotle in *JSVM* is concerned to maintain the generality of his account. He does this in two main stages. The first concerns the position of the source of the soul (chapters 1-4), where he is trying to show where this centre in plants and animals lies, and the peculiarities in insects, among other things, making them so close to plants. The second stage concerns the mechanisms of cooling (chapters 6-22), concerned largely with relatively complicated ideas about how lungs and gills in the different kinds of animals perform their tasks, in conjunction with the vital function of eating and its relation to breathing.

Now of course this kind of generality is unlikely to make a logician happy: there is no *logical* reason that there might not be other kinds of living things that Aristotle's *Histories*, which he is using here, failed to bring to light. But the idea is nonetheless attractive; it is difficult to imagine how else one would form a general idea or definition of life than by investigating in what way it is present in all living things.[38] Of course to get such an investigation off the ground one needs a preliminary grasp of what life is; and if the investigation is going to be a successful one, then the preliminary idea is not one that should at some point be overturned. In other words, the preliminary grasp is confirmed by the investigation; confirmed, not merely because nothing bears witness against it, but because it is explained. Verification is possible, because the preliminary definition is cashed out in specific terms.

Living things are divided at the most general level into plants and animals,[39] and this division is one that recurs in the attempt in chapters 1-3 to locate the seat of the soul in all living things.[40] Among animals the main division lies between those with blood and those without; the former have a heart producing blood, the later an analogue.[41] In chapters 4 and 5 where the role of heat and its need for preservation are discussed, plants are not mentioned; this is made good (at least cursorily) in chapter 6, where their heat and its cooling are discussed. The treatment of cooling among animals is determined by the observation that only animals with lungs breathe; differences in lungs are correlated to the different kinds of animal.[42] The division between water and land animals is central to the idea of cooling in that this occurs through the environment – which is sufficient for small bloodless animals without any specific organs to perform it; long-lived bloodless animals have their own mechanism using connate breath.[43] Blooded animals are divided into those with a hot bloody lung – the internally and externally viviparous – and the oviparous, usually with a spongy lung.[44] Water animals are divided into those with lungs and those with gills, the latter corresponding to footless animals.[45] Two chapters (17, 18) are devoted to organs connected with cooling, but in ways that require careful handling – the mouth, and blowholes in cetaceans. Taken as a whole, the investigations of 1-3 and 6-18 display a powerful move towards generality.

Thus the upshot of these two investigations is that all living things are hot, and that their heat is cooled. Heat and indeed cooling were introduced as being the tools of nutrition. In this way the definitions just bring together the central elements of the enquiry. But, since they are general, they abstract from all the cases reviewed, and which underlie the generality. This is clearest when we consider Aristotle's efforts to keep the account of perishing general: so-called death in animals; drying out in plants.[46]

The definitions apply to all living things, and thus to none specifically, that is, not in such a way that they apply *tout court* to any existing form of life. A central role is played by nutritive soul, that soul which occurs isolated in plants. But it is to be noted that even in plants it will occur in each case in a specific fashion. Similarly in animals it exists in a specific fashion, such as to form and sustain specific varieties of perception. Nutrition is an activity or actuality which only occurs qualified by an adverb of fashion, e.g. 'in a human manner'. For it cannot be performed, if not in some specific fashion. (This point has no implications for the question of whether activity can be understood that does not necessarily have any qualifications.) Furthermore, matter – organs – provides the basis for the ways in which living things can differ from one another in their actions and ways of life. Without matter there would be no differentiation.

Being hot does not make something alive; there is therefore a problem with the convertibility of heat with the presence of nutritive soul: it is not true that whenever heat is there, so is nutritive soul. This problem, however, does not arise because the domain has been restricted right from the start (at least of *de Anima* II) to living things. Clearly only the heat in living things is meant. This move rescues the definition from falsity; but it does not make the term 'heat' any clearer. The comparison between living things and fire is of course of very limited use and masks many difficulties. And it is a comparison which is nowhere argued to be stringent. The main treatment of heat and cold (*PA* II 2) offers no definition of heat, but is instead concerned primarily with the difficulties in saying when one thing is hotter than another. More hopeful is the specification in *GC* II 1 329b26 ff: heat brings together what is of the same kind and separates the heterogeneous, whereas the cold brings together both what is of the same kind and the heterogeneous. As we have seen, these operations are closely connected to what happens in nutrition: living things assimilate food to themselves, and so separate off from food those parts unlike themselves. But one does not have to be particularly mathematically minded to be worried by the way in which there seems to be no way to divide hot from cold, except gradually. Hot and cold may be distinguished by perception, although that is not as clear for Aristotle as we may think; but there is no defining hot as opposed to cold. Aristotle himself wavers between saying that cold has its own nature and that it is merely a lack of heat.[47] While the extremes may be clear (although I know of no texts which say what their instances are), like other continua in Aristotle they remind one most of the continua in the *Ethics*; and the connection between the ethical mean and the balance in living things is one we have already noticed.[48] But this resemblance does not reassure one that there is a way of defining one thing as hot rather than cold, or, more precisely, saying what it is about the hot that makes it hot.

The opposition between death and coming to be shows that we are concerned with death as an event, as the end of life rather than a process within it.[49] This is the reason for another striking feature of *JSVM*: Aristotle says very little about life as such, although this is part of the topic of the work.[50] This is because the actuality of the organs described is life: the nutritive soul working in the central organ, the cooling action of the lungs. Living is being seen as an actuality with events at either end that start and terminate it.

So it is clear why Aristotle spends so much time on the apparatus of cooling: he thinks there is a very close connection between the state of the apparatus and the phenomena which he is concerned to describe in *JSVM*. The picture is of one organ that determines the vitality of the living being through the first appearance of this organ – in sanguineous animals the heart, its growth, and its fading away.[51] Aristotle knew of no timing device to regulate the growth and development of living things. We can now see that the only kind of regulating device he knew was concerned with the amount of nourishment that can be concocted. In turn, that depends on how well the innate heat is being tempered. This part – the lungs or gills – itself grows and decreases in size, its size being apparently proportionate to the amount of heat and hence nourishment that the animal concocts. For Aristotle, development, in the sense of the differentiations of parts of the body, takes place in certain determined stages.[52]

Notes

Life-cycles

1. Abbreviated as *LBV* and *JSVM*. For these titles see below §3.1, where we also discuss the relation between *dA* (*de Anima*) and *PN* (*Parva Naturalia*). For my abbreviations of the titles of Aristotle's works see pp. vii-viii.

2. For the metaphysical importance of this chapter, see Buchheim 2001; and for a treatment of the question of the separate forms of living things, seen in the light of their nutritive activity, cf. Hübner 2000: esp. 191-223.

3. Cf. Sorabji 1972: 17.

4. For many aspects of his theory of soul see esp. the essays in Gotthelf 1985, Gotthelf, Lennox 1987, Devereux, Pellegrin 1990, Nussbaum, Rorty 1992, Kullmann, Föllinger 1997. Lloyd 1996: ch. 4 'The Master Cook' discusses the ambiguousness of 'concoction' (*pepsis*), which is intimately connected with nutrition, and the difficulties this gives rise to in formulating scientific theorems, as specified in *APo*. Althoff 1997 concentrates on the material aspect of nutrition in *GA*. Cf. also Matthews 1992, Code, Moravcsik 1992; Nussbaum 1978: 76 ff. Michael of Ephesus' commentary on *LBV* and *JSVM* has also proved invaluable on the details of the treatment. (On the dating of Michael (twelfth century), see Browning 1962, Mercken 1973.)

5. See Kahn 1988 on this post-Cartesian omission.

6. A good example is provided by Plato, *Phaedo* 66b-d, and contrast his Socrates' early interest in nutrition (96a1 f).

7. Freudenthal 1995, Gill 1989a.

8. Freudenthal 1995: 127-30, 140-2. Another version of vitalism is attributed to Aristotle by Feldman (1992: 39 f): life is caused by a substance, namely the soul. But it would be more Aristotelian to say that the soul is the formal aspect of a living substance, rather than a substance on its own. On vitalist interpretations of Aristotle, cf. Sorabji 1980: 170. See below §3.2 on the soul.

9. E.g. vital heat holds together the uniform parts of the body like flesh ('homoiomers') (Freudenthal 1995: 18, 8-9). Freudenthal argues (1995: 9-10, 19, 45) against both Gill 1989a and Cooper 1982 – i.e. against form (Cooper) and soul (Gill) as performing this function.

10. Freudenthal *always* uses 'substance' to mean 'material substance' (see 1995: 150 and n 2: he means substance made of the four elements). This is a legitimate use of the word 'substance' in Aristotle. (See esp. *Metaph*. VIII 4, but note that there we have great emphasis on the fact that material substance is suited to the function it serves.) But it is hardly a helpful usage when confronted with the use of matter and form as explanations in Aristotelian physics. See also Freudenthal 1995: 137, 43.

141

11. 'The physiological theory thus definitely implies that vital heat should be *the one causal factor* on which depend all these characteristics [size, type of moisture], including longevity.' Freudenthal 1995: 191 (my emphasis), cf. 187.

12. Gill 1989a: 203-5. In the bulk of the paper she is concerned with the question of substance. Much of the material that shapes her view of perishing is presented in her 1989b: ch. 7.

13. Gill 1989a: Part VI 'On Life and Death'; in her view, individuals consist of definite stuff which 'receives functional organisation and survives the destruction of the individual when these functions have been removed' (1989b: 134); cf. Gill 1989b: 213-14, 218-27.

14. Gill 1989b: 200 on *dA* I 4 416a13-18. This section of her account is largely based on the contrast she sees in *Metaph.* IX, esp. chs 1 and 8, between capacities for changes in other things and souls as capacities for change in the proximate matter of the same thing.

15. Gill 1989b: 234-5, 1989a: 203: 'The material properties that connect a composite with its simpler origins have a decisive and adverse effect on its life. The influence is adverse because the residual material properties, though accidental to what the complex is, are sufficient to identify a simpler body. For this reason an organism must exert itself to offset the natural tendency to decay into a simpler state.'

16. See Gill 1989b: 149-61, esp. 160, also 128-30, 166-7; 1989a: 204: 'Internal decay occurs because the lower material properties, although controlled in the construction of a composite, still promote behaviour consistent with their own lower character ... So composites do not live easily once they have been generated, and no external destroyer is needed to bring about their destruction. On the contrary, life is a continuous struggle against decay. Because the lower material properties offer constant resistance, the project of staying alive is one that demands considerable exertion.'

17. *JSVM* 23 478b32, cf. *PA* II 5 651b11. Plato makes Socrates describe a general theory about all coming to be and passing away in nature (95e10) in the *Phaedo*; and this theory is designed to enable him to give his theory of death.

18. Cf. Nussbaum, Putnam 1992: 46.

19. *JSVM* 24. In his natural works Aristotle does not divide human life into periods of seven years; he is, as usual, very sparing in the use of numbers (cf. *Pol.* VII 17 1336b40 ff, *Rhet.* I 1361b7-14). The idea of the life-cycle, with more or less definite periods, was deeply ingrained for the Greeks (e.g. Solon Frg. 27, Ps.-Hippocratic *peri hebdomadôn*). On the topic in general, cf. Garland 1990: Introduction: The structure of human life. Boll 1950: 168 f briefly discusses *Rhet.* II 12-14 on the idea of the three ages (I owe this reference to Hans Gottschalk). On the periods in life, see below §4.2. For the Greeks' changing cultural responses to mortality from the archaic to the early classical periods, see Sourvinou-Inwood 1995. Her thesis is that there is a change from seeing death as part of a natural cycle, where it is hateful but not feared (the *locus classicus* is *Iliad* 6.145-9), to the predicament of an individual in fear of extinction (see esp. ch. 4.298 ff); one can see this predicament reflected in Plato's *Phaedo*. Aristotle is conservative in his emphasis on the natural cycle, but of course the specifically human aspects do not form part of his treatment. He deal with the problem of happiness in relation to a whole, mortal life in *EN* I 10 – an idea which has close affinities with that of a natural life-span.

20. For Aristotle's view of the divine, see *Metaph.* XII 6-10; for a general application of immortality to soul, compare Plato *Phaedo* 106d-107a: to the disappointment of readers of the last proof for the immortality of the soul, Plato thinks that 'the divine and the form of life' anyway are fundamentally untouched by mortality. In a sense this is true of Aristotle as well: we are only dealing with finite life.

21. *EN* III 6 1115a26. The parallel treatment of life and death could be compared with that of virtue and vice, which would conform with the general idea that contraries are to be treated by the same science, for example, medicine can produce both health and sickness (e.g. *APr.* I 1 24a21, 36 48b5; *Metaph.* III 2 996a20, XI 3 1061a19; *dA* III 3 427b5, 6 430b20-26). Life and death, however, are not such contraries; one and the same human can become just from being unjust, but cannot change from living to dead, and remain human. There is no continuing subject, first of life and then of death, at least in the way Aristotle conceives them. Contrast Plato's argument from cycles between contraries for the immortality of the soul (*Phaedo* 70b-72e, esp. 71d-72a), which the young Aristotle reworked in his dialogue on the soul, the *Eudemus*, esp. Frg. 5, 7. This argument presupposes that there is no real coming to be or passing away, merely joining and parting of body and soul.

22. Feldman 1992: 20.

23. The value of death is sometimes discussed in relation to Epicurus' argument against the fear of death (when I am there death is not, and *vice versa* – Lucretius III 870-93, Epicurus: *Letter to Menoeceus* 124-6, *Rata sententia* 2). For a selection of articles, see Fischer 1993. A consequence of this view might be that murder cannot be rationally condemned.

24. Nagel 1979: 9; see Emson 1995: 166, for the use of the idea of natural life span and death in reflections about modern health-care; contrast Fischer 1993: 15. The idea of a natural life-span fits very neatly with fictional difficulties in coping with an indefinitely extended life, understood in terms of our usual projects (see Williams 1993, who, however, does not mention natural life-spans). In Aristotle's ethics, the idea is central that only a whole life may be called good; and this life should have a natural life-span (see *EN* I 10).

25. In the modern debate about death, the question of the direction of life arises because of the Epicurean argument that the time after our death should be just as much a matter of indifference as that before our birth (Lucretius III 832-42, cf. Nagel 1979: 7). For a survey of the literature up to 1993, see Fischer 1993: Introduction.

26. The main (surviving) texts are Heraclitus DK B 21, 36, 88, 126 cf. A 19, B 20, 26; cf. Anaximander DK B 1. For connections with the notion of nature, see Buchheim 1999: 17. Plato's use of the idea of life-cycles (*Phaedo* 70b-72e) is criticised by Strato of Lampsacus (translated in Hackforth 1955: 195, Gottschalk 1965: 123 f) for, among other things, requiring that the young arise from the old. Plato also plays fast and loose with the directions of life-cycles at *Statesman* 270d-e. This raises questions about the relation between the direction of life-cycles and the notion of time's direction.

27. Fischer 1993: 5.

28. See Fischer 1993: 7 on official guidelines in the United States.

29. *JSVM* 24; see below Chapter 5.

30. For modern definitions of life, see Feldman 1992: 31. One reason that nutrition is neglected here may be that non-living things (fires, crystals) might have a claim to nourish themselves. See below §3.3.

31. The cecropia moth, cited by Feldman (1992: 26 ff) as a counter-example to nutrition as a defining activity of living things.

32. Indeed, it is a stage of life that offers particular problems. For whereas the genetic control of the growth of things may be understood straightforwardly in terms of evolution, ageing seems not to be something you can select for: in the wild, it usually does not occur, and so there can be no evolutionary pressure determining how it occurs. Ageing is furthermore intimately connected to resistance to disease, and is not as tightly regulated as development.

33. Comfort 1979: 16-20, 41-4. He was writing before the recent explosion in our knowledge of timing devices in living things. As Comfort notes, ageing set off by a clock is compatible with an error accumulation theory, outlined below in the main text.

34. Brady 1979: 2-3, 57.

35. See e.g. Davies 1983: 4.2 Unprogrammed ageing.

36. Aristotle does not even refer to the water-clock. References to the *klepsydra* at *Ph.* IV 6 213a27, *dC* II 13 294b20 are not to the water clock, but to a 'water lifter acting on the syphon principle' (Ross 1936: 582); for a Peripatetic treatment see *Prob.* XVI 8. For Empedocles' use of the water lifter as a comparison for the mechanism of breathing see DK B 100 (taken from *JSVM* 13 473b9-474a6). On external timing (*GA* IV 10 778a5) see §4.2 below.

37. See Gallop 1996.

38. See esp. *SV* 1 454a26 f, 454b8, and Hintikka's study (1973) of time and modality in Aristotle, esp. ch. V 'Aristotle on the Realization of Possibilities in Time'.

39. Cf. *Ph.* II 8 199a33-b7 on mistakes in both nature and art. Death and ageing, however, are not mistakes because they are regular natural processes.

40. Cf. *Metaph.* IX 9 1051a4-21 esp. 19-21: there is no destruction in eternal beings and destruction is a bad thing: i.e. in a hierarchy of entities, perishable beings will come after eternal ones.

41. As to badness: he does not think that matter alone is bad. *Ph.* I 9 192a13-25 refers to privation, i.e. the lack of a form, as 'that which makes things bad' (cf. *GC* II 9 335b8 ff: there it becomes clear that privation rather than matter is meant in *Ph.* II 9 192b13-16: in both texts, it is privation that others have barely been able to grasp). The idea is that something is bad when it does not have the form required, i.e. in view of some end or function. A bad saw is one lacking the proper form of a saw; but, of course, this lack of a form is in the metal of the saw. For the use of privation with reference to life-cycles, see *Metaph.* V 22 1022b27-31: not having sight is only a privation if the living thing characteristically has sight at the age in question. The implication is that a blind baby rat is lacking nothing, although it naturally cannot do all that a mature rat can. Put generally, the acquisition and loss of capacity concomitant on going through a life-cycle are not to be seen as privation, and so not as bad (cf. the passage from *Ph.* I 9 cited at the start of this note).

42. *Ph.* II 2 194a27-33. Charlton's (1970) translation. I keep the Mss. reading at a30 with Charlton, i.e. without Ross' addition of *to*. The source of the quote is unknown. Cf. Charlton's comment (1970: 101): 'Processes may end in bad states such as death or disease; we cannot establish what a process is for just by observing where it ends, but must know independently what the best "end" for it is.'

43. *Ph.* II 2 194a12-15, 26-7, cf. *dA* I 1 esp. 403a5-b16; *PA* I 1 641a17-32 and Simplicius 1882: 301.30.

44. Cf. Simplicius 1882: 302.17-18: death is not the aim, but a necessary accident of such substance.

45. For Aristotle's idea that genesis is the only way of attaining the good apart from God's, see *GA* II 1 731b24-732a1, *GC* II 10 336b27-34. For God's life see *Metaph.* XII 9, cf. *EN* X 7 1177b16-1178a2.

46. Simplicius 1882: 302.10, 311.1-19 (quoting Alexander), cf. Alexander (Sharples 1983) *de Fato* 168.5, 1891: 103.31 ff, (Ruland 1976) *de Providentia* 81.5-11, (trans. Dooley 1989) *On Aristotle's Metaphysics* 1 142, n. 306. I owe these references to Bob Sharples.

47. *dC* I 10 280a10-23, esp. 280a14-15.

48. See *GC* II 11, esp. 338b5-11, b14-18, esp. Williams 1982: 208, Sorabji 1980: 148.

49. Cf. Aristotle's remarks about wholes in *Poet.* 7 1450b26-36. For the idea of beginning, middle and end, cf. *dC* I 1. The central role of maximum functioning reached in time was suggested to me by Thomas Buchheim.

2. Nature

1. *Ph.* II 7 198a29-31, and Ross 1936 ad loc. Of course, in another sense, astronomy is a part of physics, since the stars have their own nature, merely one that does not involve perishing (see e.g. *Metaph.* IX 8 1050b16-28).

2. We have already met them above in §1.2. Compare the way in which the account of nature in *Physics* II 1 prepares the way for the introduction of the causes or explanatory factors in the rest of book II.

3. *Metaph.* V 4 1014b16-20. For the importance of this chapter for the *Metaphysics* see Buchheim 2001.

4. See e.g. Charlton 1970: xvi-xvii, Balme 1992: 96.

5. For some remarks on the difficult question of the epistemological status of linguistic arguments in Aristotle see Waterlow 1982: 12, 59.

6. Cf. Ross 1924: ad 1014b17, Waterlow 1982: 62-6; Heidegger 1967: 292 ff. This conception of nature as a process (genesis) is only discussed once elsewhere *Ph.* II 1 193b12-18 (cf. Ps. Aristotle *MXG* 975b6). For a play on words which implies it cf. *GC* I 3 319a10-11, *PA* I 1 641b30, and Peck 1961 and Balme 1992 ad loc. It is of course of importance that he mentions this sense in both of his full treatments of the concept. Elsewhere, and commonly, all the stages of growth and preservation of the species in reproduction remain within the one nature (*Metaph.* VII 7 1032a15-25, 9 1034a33-b3, IX 8 1049b17-29, 1050a4-10, XII 3 1070a4-8, 21 f).

7. *Metaph.* VI 1 1025b26-1026a6.

8. These two senses of genesis are obviously closely connected to the Aristotelian view that reproduction and nutrition are closely connected (*dA* II 4 416b23-5, 415a23-b1, and below p. 21 on *GA* II 4 740b29-741a3).

9. And does not refer to so called 'primary matter', a substrate with no properties of its own underlying everything, which is not mentioned here at all; for the debate as to whether Aristotle has or needs this concept, see Charlton 1970: 129-45, Williams 1982: Appendix, Gill 1989b: Appendix (with further references). On an ultimate material nature that remains unchanged see 1014b26-35, the 'elements' of the earlier natural philosophers, in Aristotle's view (also *Ph.* II 1 193a9-28, Waterlow 1982: 55-8).

10. Aristotle illustrates 'primary matter' only with an artificial example, bronze in relation to things made of bronze (1015a9); and of course such an example does not make clear how it is related to growth. But it is reasonable to think that a living thing comes to be out of its food. Aristotle thinks that food before it becomes the thing in question is blood. See below on nutritive soul §§3.3, 4.4. Blood is, although in a thing, not a part of it (*PA* II 651a14, *JSVM* 3 469a1 f). See also Lewis 1994, section IV on blood as the matter of an animal; Freeland 1990.

11. Ross (1924 ad loc.) thinks the 'inherent starting point of growth' is meant, and, following Bonitz, suggests seed (cf. e.g. *dA* II 1 412b25-7 on seed as a body capable of living); but it is better to follow Alexander (cited by Ross), even if matter is discussed again in lines 16-35; for parallel formulations referring to matter see e.g. *Ph.* II 1 193a10, 3 194b24, *Metaph.* V 3 1014a26. The first lines of the chapter provide a resumé of all the uses of *physis*. And a seed is not a constituent of the grown thing; seed is nowhere else called the nature of the living thing.

12. See below §3.3.

13. Perhaps *kai* here should be taken as meaning 'i.e.'; it is hard to see what else the comings to be of growing things could be apart from the two senses identified above in the text; but 'growth' appears in 1015a16 in any case.

14. On growth as what a living thing does on its own, once it has been generated by its progenitor see *GA* II 1 735a13-22. *Ph.* VIII 7 260b30-261a12 argues that there must be a change prior to generation, i.e. the locomotion of the heavens (cf. *GC* II 336a31-b19, *Metaph.* XII 1071a15, 1072a10-12); but the first change that happens after generation in those things that do come about is 'alteration and growth'; in order to show that growth (change of quantity), rather than alteration (change in quality) is primary, we will have to show how it is that alteration takes place in such a way that growth happens, and that what then happens really is growth rather than merely alteration. See below §3.3.

15. For growth as a primary change cf. *Ph.* VIII 7 260b32: alteration is also mentioned, since, as we will see, growth does not take place without alteration. Locomotion is not the primary change for things that come to be (they can only change places when they have grown in the requisite way), although it is primary in the general scheme of things since it causes coming to be and passing away. Cf. *Ph.* II 8 199b15-17, *GA* II 1 735a12-22.

16. *Ph.* II 1 192b13-16 (cf. *dA* II 1 412b15-17). Cf. Waterlow 1982: ch. I, Gill 1989b: 218-22. Note the examples in *Ph.* II 1 – alongside locomotion, increase in size (*auxêsis*) and wasting, and alteration. In *Ph.* VIII he draws the distinction between those changes whose principle is in the thing itself (locomotion), and those natural changes in the wider sense of those that are there as soon as *physis* is present in the thing. The point is that things cannot be responsible for their own inception, so their *physis* has to come from without; then it has a *physis* and can develop and nourish itself.

17. *MA* 5 700a34 ff, *GA* II 1 734b14, *Metaph.* VII 7 1032a15-20, IX 8 1049b24-5, 1050a5-10. This thought is obviously connected with the idea that a form existing prior to the living thing to be brought about can direct the achievement of form by the new living thing. See below on the problem that the final cause or form when not yet there cannot be a cause.

18. For an alternative interpretation of rest here, see Waterlow 1982: 121-2, 128. She takes rest to be remaining in a state, as opposed to changing to that state. On my reading, rest is just the end of change, and does not include the activity required for stability sc. at the point reached by the change.

19. *Metaph.* V 4 1014b20-6.

20. Cf. Kirwan 1971: 130, who thinks the problem Aristotle tackles here is why growth is self-change, if it happens because of food, that is, an external factor:

Aristotle's answer is that the food, when has been changed into the thing fed, has the same form as that thing.

21. There are difficulties with the concept of 'growing together' (*symphysis*): it can mean abnormal growth (e.g. Siamese twins, *GA* IV 4 773a4 ff) as well as normal growth. An arm does not come to be my arm by growing onto me. Rather, as part of my growth, my arm grows; we grow, as it were, together. Cf. *Metaph.* VII 16 1040b15 and Ross 1924 ad loc. with IX 1 1046a28-9: insofar as something has grown together it does not undergo any change under the influence of itself, since then it is one thing, and not another (cf. *Ph.* IV 5 212b31). In *Metaph.* V 4, the growth meant is normal; the contrast is with the stage when the matter has not yet become part of the living thing. This change is an essential element in the analysis of nutrition (see below §3.3). When they have grown together, they are one thing – this is also the point of XII 3 1070a10-13: *symphysis* is used of matter which has not grown together and become a *physis*, so must refer to normal growth; since we are dealing with nature in V 4, it is reasonable to think that we are dealing with normal growing together. This is anyway presumably prior to any abnormal developments (at least if some form of normal growth is necessary for the continued development of a thing that functions i.e. can nourish itself).

22. *Metaph.* X 1 1052a22-5; XI 12 1069a5-12.

23. Cf. *PA* I 1 640b29-641a17 against Democritus.

24. Keeping *tôi ex archês meizôn de hautê estin* with Drossaart Lulofs, against Peck. Presumably, the capacity to work on food of a newly formed embryo is less (relative to quantity) than that of the fully grown animal.

25. *GA* II 4 740b29-741a3.

26. As argued above in §1.3.

27. If growth and nutrition are connected, then we have here a first step towards making more plausible the use of the etymological connection between growth and nature with which Aristotle begins his treatment of *physis* in *Metaph.* V 4.

28. Why does Aristotle think that the matter in both cases is the same? His view of blood is crucial here (cf. *JSVM* 3 469a1, *PA* II 651a14, *GA* I 20 729a32, II 1 733b26). Blood is the proximate matter from which living things are formed and *katamênia*, the matter for the formation of the embryo, is formed of blood, sc. of the mother. On the relation between milk and menses see *GA* II 4 739b25, IV 8 777a3-8. *Metaph.* V 4 1014b20-6. On nutrition see below §3.3, and on blood §4.7.

29. See above p. 18.

30. *Metaph.* V 4 1014b35-1015a11. Lines 1014b37-1015a2 contain a quotation of Empedocles (DK B 8, lines 1, 3,4), which I leave out. Lines 1015a7-11, on two senses of primary matter, have been discussed in the last section above.

31. It is presumably to account for the wider use of 'nature' that Aristotle says that all *ousia* (substance) is said to be nature metaphorically, since nature is one kind of *ousia* (1015a11-13).

32. See *Ph.* II 1 193b16-18.

33. See the parallel in *Ph.* II 1 193a31-b8.

34. For the identity of the principle of coming to be and its end cf. *Metaph.* IX 8 1050a6-10.

35. Waterlow (1982) argues that the form must be a capacity for bringing forth a thing of a certain kind: for only such a capacity is common to both progenitor and offspring; the problem with this view is that Aristotle thinks form is actuality (see especially *Metaph.* IX 8), rather than capacity.

36. See esp. *GA* II 1 734b5-735a26.

37. In spontaneous generation, the co-operation of the parts takes place in such a way *as though* there were an organising form, which, in these cases, there is not. See *Ph.* II 6, *Metaph.* VII 9.

38. Far from denying that matter (and so the female) has a dynamic role to play in generation, Aristotle's view requires the co-operation of the capacities for change in the matter provided by the female. See esp. *GA* II 4. Cf. Föllinger 1996: 224 ff.

39. See below Chapter 5 for some remarks on the definition of youth.

40. *MA* 5 700a34 ff, *GA* II 1 734b13 f, *Metaph.* VII 7 1032a15-20, IX 8 1049b24-25, 1050a5-10.

41. Sorabji 1980: 40-2, Frede 1980.

42. *dA* II 4 415b21-8.

43. *Metaph.* XII 3 1070a21-4.

44. Cf. Woodfield 1976: 133 for an approach to the problem of final causes working backwards in time, based on the idea that teleological descriptions possess no temporal index.

45. For these two aspects of living form contrast *Ph.* II 7 198a24-6 with *GC* I 7 324b15 on form not being a moving cause.

46. *Metaph.* V 4 1015a13-19.

47. Cf. *Metaph.* IX 7 esp. 1049a15-17, and Gill 1989b: 227 ff.

48. Cf. Charlton's remarks on the argument for teleology *Ph.* II 8 198b34-199a8 (1970: 123): 'The growth of a living thing is such a process [sc. a continuous change towards an end for the sake of which it takes place]: living things do not come to be at random, but from definite kinds of seed ...; so growth is for the mature plant or animal and the performance of its function.' The end must also be mentioned (cf. *Metaph.* IX 8 1050a4-10).

49. For this argument for form cf. *Metaph.* VII 8 1033a32-b10, cf. below §3.4.

50. The first thing to notice about Aristotle's view of changes is that these themselves are end-related: what the change is, is determined by the end of the change. A change in a quality ends with some quality, a change in quantity ends in some particular size. Thus, if form is to be understood as a complex of changes, then at least it will not be a complex of wholly indeterminate entities, but of ones essentially related to ends. For the canonical account see *Ph.* III 1, also V 1. Since changes are defined as activities that have not reached their end, it might seem that we have the same problem with changes as we do with growing things: how they can be determined by something not there. On changes as aimed at their own termination and the preservation of substance, see Kosman 1969, Waterlow 1982: ch. III.

51. *Metaph.* VI 1, VII 10. Cf. Anscombe 1979. Frede 1990. Frede/Patzig 1988: I 45; Nussbaum/Putnam 1992: 29.

52. Cf. Gill 1989b: 233.

53. On the analogy between art and nature, cf. Balme 1992: 94 f, Broadie 1990.

54. *Ph.* II 1 193a31-6. Aristotle (*Ph.* II 8 199a12) thinks that if artefacts did grow, they would come about in precisely the same order as they actually do at the hands of their maker. From the status of both growth and artificial production as ordered processes Aristotle argues for the presence of ends in nature *Ph.* II 8 199a8-20. For this use of 'not yet' cf. *Metaph.* IX 8 1049b19-23.

55. See p. 21.

56. *GA* II 4 740b25-9, cf. also *Meteor.* IV 3 381b3 ff, where the parallel is applied especially to concoction (cf. Althoff 1992: 184 f).

57. On the question of reflection – possibly a major difference between nature and art – see Broadie 1990; she points out that Aristotle's *technê* is idealised, and does not reflect on ends, but is defined by them. On the lack of reflection in *technê* cf. *Ph.* II 8 199b26, Kullmann 1979: 50.

58. Quoted above, p. 21.

59. *Metaph.* IX 2, on which see King 1998.

60. I translate *dia tês kinêseôs autôn* by 'through the changes caused by the tools', rather than 'through the movements of the instruments' since the point is that the tools by changing or moving cause changes in the thing in production, rather than that the tools are just moved or changed. 'Change' rather than just 'movement' since Aristotle has heating and cooling in mind in nature; and a smith's fire or a cook's oven are good analogues.

61. Cf. also *GA* II 1 735a3-4.

62. Cf. *Metaph.* VII 7, 8, IX 8, *PA* I 1 640a23-6.

63. See Burnyeat et al. 1979: 37, Nussbaum/Putnam 1992: 32.

64. On the relation between Aristotle's conception of change and that of substance in the *Physics* see Waterlow 1982: ch. III. The main point is whether the regulation of change is independent of his conception of substance, or purely subordinated to it.

65. Burnyeat 1992: 26, Ackrill 1972.

66. Another way in which Aristotle sees the separation of form and matter in natural things lies in his analysis of generation: the male provides the form, the female provides the matter (*GA* II 4 738b20-7).

67. E.g. Kant 1902: 294-5.

68. Lloyd 1978: 233.

69. *JSVM* 9 471b23-9.

70. On the long history of the refutation of Aristotle on the subject of respiration cf. Ogle 1897: Introduction; Furley 1984.

71. Cf. Woodfield 1976: 133.

72. On this sense of 'necessary' cf. *Metaph.* V. 5 1015a20-2: 'The necessary is that without which one cannot live, as a co-cause, e.g. food or breathing are necessary to an animal, for it is impossible to exist without these.' On necessity in Aristotle's explanations, see below pp. 82, 114 f.

73. For another general structural comparison between the relation of a flute player's hands to his instrument and natural organisation, see *JSVM* 4 469b1 ff.

74. Althoff (1992: 160 n 22) points out that modern scientists are not above comparing lungs to bellows, at least for educational purposes.

75. On these two models of teleology in *Ph.* II 8, see Charles 1991: esp. 102, 125-7.

76. Nussbaum (1978 e.g. 85) emphasises the role of sensitivity in natural things; but there is no room for change in the form of things for Aristotle, merely a variety of ways for any life form as to how it reaches its ends. Choice should not be taken as analogous to flexibility; the point about choice is that it operates in the sphere of activity.

77. Kullmann 1979: 41 notes the heuristic function of teleology in Aristotle, and compares his use to Kant's and that in modern science, noting that in Aristotle it is not merely a regulatory principle. However, he claims that it is less exact for

Aristotle than scientific theorems; and I see no reason to follow him there. Another crucial difference is that ends for Kant are always purposes (Zwecke).

78. Famously, Kant's claim is that we can approach the field of nature *as if* it had been arranged by an understanding other than ours in such a way as to be amenable to our understanding (1902: XXVII). There is here merely a verbal correspondence to the phrase Aristotle uses at *Ph.* II 8 198b28-31 to describe the position Empedocles' materialism would, in Aristotle's view, force him into. On this passage contrast Wieland 1962: 260; for 'as if' of purposeful design cf. *dC* II 9 291a24-6.

79. E.g. *GC* II 10 336b27; good examples of use of the maxim as heuristic principles in *IA* 2 704b12-18, 8 708a9-20; *GA* V 8 788b20-4. For another interpretation of these passages, see Lennox 1997. One question about such possibilities is how they are related to Aristotle's apparent commitment to the idea that a possibility must, to deserve the name, be realised (*Metaph.* IX 4).

3. Soul

1. For the difficult question of the relation between these two treatises and the great zoological works, see below nn 10, 12, 20, 21, 38.

2. *dA* I 1 402a4-10. The definition of nature is repeated in *dA* II 1 412b16 f, shortly after the definition of the soul.

3. The possible exception to this rule is so-called 'productive' thought (*dA* III 5).

4. *logoi enuloi* 403a25. 'Attributes': The Greek has *pathê*, which can mean 'affection, emotion'; and emotions are being discussed. But clearly there is no reason to restrict *pathê* to emotions: they include everything that can happen to the soul; hence the translation 'attribute'. See Hicks 1907: ad 403a16.

5. *Ph.* II 2. The student of nature studies such soul as requires matter: *Metaph.* VI 1 1026a5-6; cf. also *PA* I 1 641b8-33.

6. *Ph.* II 2 194a12-15, 26-7; Simplicius 1882: 301.30

7. *Ph.* II 9 200a32-4.

8. One central part of the definition of soul is actuality (*entelecheia*), a concept which does not appear in *PN* (Kullmann 1979: 22); but once it has served its purpose of setting up the framework to be filled, there is no need for the concept itself to reappear. Note, however, the use of nutritive soul in the definitions at the end of *JSVM* (below Chapter 5).

9. Cf. Alexander's remark (1901: 5.13-19) on the phrase 'things common to body and soul' (*SS* 1 436a7-8): they are those peculiar and common to different kinds of animals and are to be investigated by looking at the parts of the body 'in which and through which the activities (*energeiai*) of things with soul [exist]'. On the necessity of investigating both body and soul to understand the accidents of soul: see Hicks 1907: ad 402a9: '[An attribute of soul which through the soul also belongs to the animal] is the normal type of attributes of soul, whether active operations or passive states. As expressed below, 403a3-10, the body as well as the soul shares in them, and therefore their definition ought to take account of the body (403a16-27), and psychology becomes a branch of physics (403a27-b7); in other words this second class of attributes or "states" of soul are *achorista tês physikês hylês tôn zôiôn* (403b17).'

10. Siwek 1963: IX. Jakob Freudenthal 1869: 84 remarks that *PN* solves the aporia posed at the start of *dA*, referred to above in the main text: which affections (Freudenthal adds: and activities) of the soul belong only to the soul and which belong to both soul and body. It would be more accurate to say that *PN* assumes that the actions it deals with are common to body and soul; an important paper attacking the view of Nuyens (whom Ross largely followed 1955a: 3 ff), according to which the entelechy view of *dA* is superseded by the instrumental view in *PN*, was Kahn 1966. Alexander was in no doubt that *PN* and *dA* belong together: what has been said about the soul is the basis of what is going to be said (1901: 4.8-18). Modern readers who treat the two works together include Kahn 1966, Sorabji 1972, Modrak 1987, Lloyd 1992, Everson 1997, Johansen 1998. Some interpreters have doubts about the unity of *PN* (Wiesner 1978: 241), but, as Siwek (1963: X) points out, *SS* 1 does mark out a subject which is then treated in the treatises: the principle operations that are common to body and soul. Here is not the place for a discussion of the relation between *PN* and the great treatises on animals (*HA*, *PA*, *GA*); for a cursory sketch of two schemes, one of which involves the great treatises, see Ross 1955a: 2; for a careful formulation of the relation between Aristotle's psychology and zoology see Lloyd 1992: 147: the psychology is a 'major articulating framework' for zoology – but it leaves room for supplementary material.

    11. *SS* 1 436a1-8, esp. 436a5.

    12. *SS* 1 436a4-5, 7-8. The second is suggested by Jakob Freudenthal 1869: 81-2, with the addition of 'functions' (*erga*) presumably from *dA* III 10 433b19 f where the phrase is used to refer forward to *PN* (and because of the common actions *praxeis*; cf. also *PA* I 5 645b14-20, which he cites as being closely connected with the project of *PN*, cf. also *PA* I 3 643a35). The absence of any talk of functions in Aristotle's formulation in *SS* 1 is warranted by phenomena, which, while involving both body and soul, and comprehensible in a teleological system, themselves serve no ends, like dreams. On the title *PN* and its inadequacies see also Siwek 1963: ix. According to Freudenthal, it goes back to Thomas Aquinas' pupil Gilles de Rome (thirteenth century). On the subject of *PN* cf. Lloyd 1978: 230. With the important exception of *dA* III 10 433b19, already referred to, Aristotle himself cites *PN* by the individual treatise names (Freudenthal 1869: 81) (e.g. *GA* V 1 779b22): but as Freudenthal notes, this gives us no licence to think that *PN* have been merely stuck together with no unifying topic in mind. Freudenthal sees this topic as being the function of organs (with reference to *PA* I 5 645b14-28). Kahn (1966: 20-1) speaks of psychology in *dA* and physiology in *PN*. Siwek (1963: ix) says that for Aristotle the whole *PN* is psychology, but for us *SS*, *MR*, *SV*, *Ins.* and *DS* are psychology, and *LBV*, *JSVM* and *dR* are biology. Ross' (1955a: 16) account was dominated by Nuyens', now widely rejected assumption of the incompatibility of soul being actuality and soul being in the heart. Van der Eijk (1994: 68-9 n 62) points out that the division of *PN* into psychology (*SS*, *MR*, *SV*, *Ins.*, *DS*) and biology (*LBV*, *JSVM*) is anachronistic: the whole work is psychology. Van der Eijk thinks *PN* is on the affects and processes, i.e. those functions an ensouled being is capable of, and for which the structures described in *PA* and *HA* exist. He thinks further that *dA* describes the parts and functions, whereas *PN* describes the processes. This underestimates the role of physiology in *JSVM*, allowing a description of the various ways the actions are performed, and the general account of the actions this ultimately provides.

    13. Especially *dA* II 6-11 on the senses. Cf. Johansen 1998: passim.

    14. Balme (1987c: 10), who does not expressly see *PN* in the context of *dA*, thinks that the 'biological' parts of *PN* are about vital heat, more specifically, the question how, 'indeed why' animals differ, which remains a question, given that we like Aristotle think the function of animals is to survive and reproduce. Even if we

note that plants should be included here (*LBV* 6, *JSVM* 6), and that Aristotle gives no reason here for the variety of living things, Balme has pointed to the central tension between a general account of life (soul) and the varieties in which this exists. For physiological detail, *PN* refers to e.g. *HA* IV 1-3 at 477a5, *HA* I 17 496a27-34 at 478a28, II 17 507a2 at 478b1. As has been noted, the order we are presented with is a reading order; even so it is not without its ambiguities, when we consider the references to the great zoological treatises: 468b31 ff refers to *PA* III 4 665b15-16, 5 668a4-9, but *PA* III 6 669a3 refers to *JSVM* 16, 22: fish have no lungs but gills as has been said 'in the treatise on breathing' (*en tois peri anapnoês*). *PA* IV 13 696b2, 697a22 refers back to *JSVM* 16, 19. *SV* 2 455b28 is referred to at *PA* II 7 653a19-21 in the perfect 'it has been said'. Finding a reliable criterion to identify 'spurious' references is difficult (cf. Thielscher 1948). Some, like the last three cited can be easily excised from their surroundings; but that alone is not enough reason to do this.

15. There are the following references back to the *dA* from the *PN*: *SS* 1 436a1, 436b10, 3 439a8, 16, 18, 440b28; *MR* 1 449b30; *SV* 2 455a8, 24; *Ins.* 1 459a15; *JSVM* 1 467b13, 14, 474b11. Forward references to the *PN*: esp. III 9 432b9-13; cf. also 10 433b19-21.

16. *SS* 1 436a1-17.

17. *ta megista ... phainetai ... koina tês psychês kai tou sômatos* 436a7-8. Negatively, this may be taken to mean that *PN* says nothing about that part of the soul that is separable; to that degree, *PN* is more truly a work of natural philosophy than *dA* (cf. *PA* I 1 641a17-b10).

18. 436a12-15. 'Conjunctions', *syzygiai*, appear also in *Phaedo* (71c9) in the argument from cycles for the immortality of the soul, the point being there that all changes runs between pairs (*syzygiai*), such as sleeping and waking, so that any such change presupposes the continued existence of the changing thing. This use is connected to *PN*'s concerns, but is of course not entirely parallel: Aristotle does not think that life and death have to be treated together because what is now alive used to exist in a state of being dead. Indeed, his use of the term *syzygiai* seems to be to avoid using the idea of opposites, so essential to his analysis of change. Nonetheless the two poles are only comprehensible together; and change is still important, since both phenomena in *PN* are comprehensible on the basis of natural changes (the argument from cycles is expressly to prevent *nature* being 'lame', i.e. lopsided 71e9).

19. *JSVM* 23 478b22, below Ch. 5.

20. The common actions of living things play an important role in the methodological deliberations in *PA* I. See Balme 1992: 116. On 'common to many' cf. *PA* I 1 639a19-22, 3 643a8, 5 645b6-10, 14-20. The last text is important because, while *PA* is a work about the parts of animals, the methodological interest in parts does not remove whole living organisms from their central position.

21. As for *orexis* 'striving', the *dA* refers forward to an account at III 10 433b19-30. See Nussbaum 1978: 9. Siwek (1963: xi with n 19) points out that a wide variety of Mss place *MA* after *DS* and before *LBV*, although he himself follows the tradition in not including the treatise in the *PN*. Nussbaum sees *MA* in a long series of lectures on biology, which ran as follows: *PA, IA, dA, SS, MR, SV, DS, MA, GA* (Nussbaum 1978: 12 following Ross 1955a: 2, Jaeger 1913: 31 ff). (However, given the references to *LBV* and *JSVM* in the rest of *PN*, they should not be excluded from this list: they are referred to in the introduction to the whole *PN* at *SS* 1 436a14-15). Nussbaum (1978: 9) further thinks that *dA* III 10 433b19-30 refers to *MA* and not to *IA*, and thinks that *SS* 1 436a9, referring to *orexis* 'striving', 'associates' *MA* with *PN*. Yet *SS* 1 serves as an excellent introduction to the whole of *PN* including *MA*. Thus the latter could very well come after *DS*. *MA* 10 704a3 f makes clear that *SS*, *SV*, *MR* precede *MA* (also that *GA* is to follow *MA*). Jaeger includes *MA* in *PN*: *MA* is about the general reasons for motion (698a4) just as the *PN* deals with perception, waking, etc. not as it appears in each kind, but as something common to all (1913: 39-40 and 38, n 2). He also points out (1913: 35-6) that motion, like perception, requires body and soul. *PA* I 1 639a19-22 lists sleep, growth, wasting, death and other 'affections and conditions' (*pathê kai diatheseis*) of this sort which he says he is not in a position to speak of 'with clarity and precision'. This would seem to be the task of *PN*. At *PA* I 5 645b33-5 he talks of the common affections and actions of living things (*pathê kai praxeis*). Cf. *PA* I 1 639a19-22. For the connection between *pathê* and *praxeis* see Bonitz 1870: 556a53; *pathos* is used of perception e.g. *MR* 1 450b5, *dA* II 12 424b3 ff. One does not need to go as far as Bonitz and say that *pathos* and *praxis* are synonyms; they can be used of the same phenomena insofar as these are actions involving, making use of some affection.

22. This passage in *SS* 1 436b1-6 has many verbal similarities with the account of memory in Plato's *Philebus* esp. 34a3-5, 10, b6-8.

23. G.R.T. Ross thinks that *praxis* usually refers to 'distinctively human actions into which deliberation and thought enter' (Ross 1906: 123-4 ad 436a5); he cites e.g. *EN* VI 2 1139a31.

24. *EE* II 6 1222b19, 8 1224a29, *EN* VI 2 1139a20.

25. *PA* I 5 645b14-20. Cf. the systematic use of *bios*, life, alongside action, parts and character, of animals in *HA* (I 1 487a11). For humans one's (way of) life is determined by choice like one's actions; other living things have both, but without the choice. Cf. Keyt 1989.

26. *PA* II 1 646b11-19. This is clear also from 645b28-33. The latter passage might also suggest that the reason such actions are called 'actions', at least in *PA* I, is that they, although without the representation of ends, do attains ends; cf. above §2.3.

27. Alexander of Aphrodisias (1901: 4.3-6) on 436a4 notes that *praxis* here is used more widely than strict usage would allow: strictly *energeiai* are fitted to no non-rational being.

28. The texts for this are *Metaph.* IX 6 (1048b18-35), *EN* X 4, *Ph.* VIII 3. Cf. also *SE* 22 178a9. *EN* X 4 on pleasure provides the best parallel to IX 6 1048b18-35.

29. On *praxis* and life see the texts collected by Bonitz 1870: 631a36-44.

30. For the purposes of this speculation, I ignore the fact that *kinêsis* is used in the title of *MA* in the narrow sense of locomotion (Nussbaum 1978: 274); it can also serve as a catch-all term for change (*Ph.* III 1 200b14, V 2 225b10-226a26).

31. as no. 48. The title is given (with variants) by the Mss. (Ross 1955a: 2-3, Moraux 1951: 297, ch. VI on Ptolemy's catalogue). The start of *LBV* (464b31-465a1) refers forward to a work on life and death i.e. *JSVM* and a work on health and sickness. The end of *LBV* refers to *JSVM*, i.e. a work on youth, old age, life and death (*LBV* 6 467b7). Some of the other treatises in *PN* are very closely connected with one another, emphasising the unity of the work, e.g. *SV*, *Ins.*, *DS*. Dreams are treated subordinately to sleeping (Van der Eijk 1994: 73). Thielscher seems undecided on whether there is a close connection between the treatises of the *PN* or whether they were only put together to form a work at some late date (1948: 244 and 246). The treatises refer to one another as follows: at the end of *SS* to *MR*; *MR* has no reference to another work at its end. *SV* 2 456a10 ff refers to what 'will be said' in *JSVM*; 456a27 refers to *Ins.*. The end of *Ins.* has no reference forwards. The end of *DS* (464b17) refers back to *SV*, *Ins.* and *DS*.

32. Michael of Ephesus 1903: 87.3ff.

33. Bekker 1831. He has *peri zoês kai thanatou* as the heading for his pp. 469-70, containing part of *JSVM* 3 and the whole of chs 4 and 6. There is some justification for this in the subject of chs 4 and 5 (see below §4.4), but no break is marked in the text, nor, apparently, by any Ms. Siwek (1963: xi) takes *JSVM* as a single work, although he divides it into sections in a very idiosyncratic way (467b10-470b5: *de Juventute et Senectute*; 470b6-478b21 *de Respiratione*; 478b22-480b30 *de Vita et Morte*).

34. Ross 1955a: 2, who, after noting the unity of the text 467b10-480b30, suggests following Biehl in calling the treatise *de Juventute, Senectute, Vita et Morte*; Michael marks no new treatise with *de Respiratione* (1903: 112.5). As Ross further remarks, *gar* 'for' at 470b6 makes it impossible to see a new treatise beginning here.

35. The forward reference to *JSVM* (*LBV* 6 467b7-9) does not mention breathing as a subject.

36. *JSVM* 1 467b12-13. 'Inhaling and exhaling' are, however, also listed as among the four most important pairs at *SS* 1 436a14 f.

37. Moraux 1951: 297. Van der Eijk (1994: 72 n 71) sees these pages as forming one treatise. Cf. Ogle (1897: V): 'there seems ... no adequate reason for any subdivision whatsoever of the treatise and it appears more consistent with its internal structure to regard it as a single work dealing with several closely connected topics.' In my references to chapters in *JSVM* I leave out the *dR*; thus *dR* 1 = *JSVM* 7 (following Ross, not Siwek 1963, Hett 1936 or Ogle 1897).

38. as his no. 47, Ross 1955a: 3. This would seem to include that part otherwise known as *de Respiratione*, which Ptolemy does not mention. Bob Sharples tells me that in the glossary attributed to Alexander in MS Ambrosianus Q 74 sup, *dR* is separated from the rest of *PN* by *PA* and *GA*, and this is taken to be the order established by Andronicus of Rhodes, removing *dR* from its original place in *PN*. (Siwek (1963) prints 478b22-480b30 as a separate treatise *de Vita et Morte*, presumably because life and death are defined there; but then so are youth and old age; and breathing plays an important role in all these considerations.) *LBV* 1 464b32 refers only to life and death as subjects for the remaining treatise, alongside health and sickness. One passage of importance (*JSVM* 27 480b21-30) right at the end of *JSVM*, picks up the topic of health and sickness mentioned at *LBV* 1 464b26-30, b32-465a1. They are part of the work of the *physikos*. No such treatise survives. The closing remark of *LBV*, which introduces *JSVM*, suggests that *JSVM* concludes the course on nature. Thus it would appear that a work on health and sickness should be included in physics and if *JSVM* finishes the work on physics, it should come before its end. On health see below pp. 59 (n 144), 89 f.

39. The physiological aspect of the treatise is well dealt with by Ogle in the introduction to his translation (Ogle 1897).

40. *tôn syngenôn tautês tês skepseôs* 480b21f.

41. *JSVM* 467b13 f, cf. 14 474b11.

42. *dA* II 2 414a20-2, cf. the context 14-28.

43. Cf. e.g. Ross 1955a: 7 on Nuyens 1948.

44. The heart is most in need of cooling (II 8 420b20-1, 25-6 with a reference to the later treatment). The heart is associated with the emotions (I 1 403a31-b1,

4 408b7-9, cf. 408b23-7); for a fuller account see Tracy 1983: 325-7. On the heart as a principle of sanguineous animals, cf. *PA* III 4 666a33-5. Tracy also points to *Metaph.* VII 10 (1035b14-1036b25) as a text in which both conceptions of soul appear. Aristotle is apparently indebted to the Hippocratic *On the Heart*, especially on the heart as the organ responsible for life: natural (*emphyton*) warmth is in the heart (ch. 6 IX 82 Littré, ch. 12 IX 90, 92 Littré), which is mainly cooled by breathing (IX 86 Littré); in the Hippocratic *On Flesh* the most warmth is in the heart and blood vessels, where it is, however, associated with pneuma (VIII 592 Littré).

45. Alexander of Aphrodisias (1901: 2.10-24) in his introductory remarks on *SS* notes that saying that the common activities (*energeiai*) of animals and things with life are common to body and soul provides support for the definition of soul as the actuality of a physical organic body. Even if one does not think that the definition of soul gains support from activities being shared by body and soul, there is every reason to agree that there is a close connection between *dA* II 1 and *SS* 1.

46. Cf. *dA* II 3 414b33-415a3, 2 413a15, b9-10.

47. Body is not one of things that are said of something, rather something of which other things are said, and so is not soul (412a17-19). Aristotle's ontology is in part based on his view of statements.

48. Cf. Alexander of Aphrodisias *Mantissa* (Bruns 1887) 104.11, *de Anima* 16.12-18, cited by Hicks 1907: 311 ad 412a20.

49. Cf. Hicks (1907: 314 ad 412b4): 'we are not so much laying down the nature of the soul as indicating the scope of the enquiry.'

50. See the views in the literature collected by Hübner 1999: 1 n 1, 4 n 5. The whole of this section is much indebted to Hübner's paper. For another interpretation of the importance of activity in Aristotle's ontology, see Kosman 1984, 1994.

51. Hübner 1999: 22-7 argues that in this comparison knowledge is an actuality on the idea that knowledge must be realised in the metabolically maintained organism, especially the images needed for thought (cf. *dA* III 3, 7, *MR* 1). This is an extremely contentious field; and I am not sure that saying that something is realised in matter is the same as saying that it is actual; capacities presumably are in matter as well.

52. For learning cf. *EN* II 1 esp. 1103a34, *Metaph.* IX 8 1049b29-1050a3, Burnyeat 1980: 73. Aristotle's view is that in learning to do something, one learns through activity; this confirms that in *dA* II 1 knowledge is not merely capacity.

53. So Hicks 1907: 313 ad 412a27: 'the earliest in development. Not only does this meaning directly follow from the words *protera têi genesei* just before [a26], but it agrees with Aristotle's intention to discover the most comprehensive definition (*koinotatos logos*), that is, one applicable to soul even in its simplest, least advanced stage.' Hübner (1999: 24) also follows Hicks on this point. Cf. *GA* II 3 736a32-b5, b8-13.

54. Hicks 1907: 313 ad 412a27 cites *Metaph.* IX 8 1050a21-3.

55. E.g. *dynamis*. *dA* II 1 413a1, 2 413a26; disposition (*hexis*) II 5 417b16. On the basis of being actuality, the soul also is or has capacities (cf. II 3 414a25 ff).

56. Cf. Matthews 1992: esp. 187.

57. *dA* II 4 415b13.

58. *dA* II 1 412a14-15, the fuller list: II 2 413a22, cf. also I 5 411a30, III 12 434a24. For the soul and life cf. *dA* II 4 415b8-14.

59. Cf. and contrast Balme's view of the soul (1992: 90): 'The soul is the working of the muscles, the perceiving through sense organs, the nourishing and growing'; 94: '[Soul] is the animal's activity, within which the activity of the flesh is part of the animal's activity, and within that again the activity of hot and cold is part of the activity of the flesh.'

60. II 413a32 ff.

61. In *SV*. Cf. Hübner 1999: 5f, 24: the contrast between sleep and waking here is not merely an example of first actuality, second actuality, it is a phenomenon the definition must cope with.

62. Hübner 1999: 16: in the original, 'organische Selbsterhaltung'.

63. E.g. *dA* II 3, 414a31, 4 415a23-6, 416a19.

64. 414a29-32, 433b2, 415a16-25; it is relative to the assimilation of food 416a19, causes growth 434a24-6, *GA* II 4 740b29, 757b14-18, and is involved in reproduction *dA* II 4 416b15.

65. On this assumption, see the next section §3.3.

66. Cf. *dA* II 3 414b20-5.

67. See *Metaph.* IX 6.

68. Capacities are always defined in relation to their realisation; *Metaph.* IX 1, V 12. Further specifications are also necessary: things that are capable are capable of something, at a time, and in a certain way, along with other, unnamed specifications (*Metaph.* IX 5 1047b35 f cf. also 3 1047a24-6, and Plato's remarks *Republic* V 477c-d).

69. Another source of unclarity might be the relation between actuality (*entelecheia*) and activity (*energeia*); Hübner (1999: 23) thinks they are interchangeable – cf. the way both terms are used at *dA* II 5, *Metaph.* IX 1, and soul is called an *energeia* at *Metaph.* VIII 3 1043a35; and the use of both in the definitions of change *Ph.* III 1, *Metaph.* XI (this last is only a problem if *Metaph.* XI is by Aristotle). For the position adopted here, I need take no stand on this question; but I think one should be worried about eliding distinct Aristotelian terms, however close they may seem to be. The one text when he explicitly relates the one term to the other (*Metaph.* IX 8 1050a21-3, cf. also 3 1047a30 f) is not immediately helpful.

70. See Ackrill 1972: 69-70, Burnyeat 1992: 17, Freeland 1990.

71. *dA* II 1 412b25-6.

72. On the compresence of actuality / activity and capacity in living things, cf. *dA* II 5 417b2-5 (the exercise of the capacity does not destroy it) cf. Alexander of Aphrodisias, *de Anima cum Mantissa* (Bruns 1887) 104.11, Hicks 1907: 311 on 412a20.

73. Mansion 1978: 12, followed by Hübner 1999: 26.

74. On this analogy, cf. Burnyeat 1992: 17.

75. Contrast the following remark by Gill: 'Since the proximate matter derives its own essential nature from the form, the proximate matter makes no independent contribution to what the composite is' (1989a: 201-2, cf. 202-3).

76. On capacities in *PA* see e.g. *PA* II 1 646a14, cf. 646b10-19.

77. Cf. Ross 1961: ad *dA* II 4 415a18-20, 'If we are to understand what the faculty of "doing so and so" is, we must understand what "doing so and so" is.' On the analogy between form and matter on the one hand and on the other, activity and capacity cf. *Metaph.* IX 6 1048a30-b9. On the identification of form with activity cf. *Metaph.* IX 8 1050b2.

78. *Metaph.* IX 6 1048a27, a30-2, a35-b9.

79. Cf. Hamlyn 1993: 101 ad 417a21: he refers to his notes on 412a6, 412a22, Hicks 1907: 354 on the same line refers to 412a10, 22.

80. It goes back to Philoponus 1897a: 204.10; for use and discussion of the scheme see e.g. Johansen 1998: 26-7, 269-72.

81. 417a21-b2.

82. 417b29-418a1 cf. 417a31-b2.

83. Trans. Hicks, emphasis added. See also III 12 434a23, *GC* I 5 322b24.

84. Cf. *dA* II 3 414b29.

85. On the status of the definition as general, cf. 412b4; the preliminary definition of life: 412a14-15; and the need for explanation in a definition II 2 esp. 413a15.

86. *dA* II 2 413b5-7, cf. I 5 411b27-30, II 2 413a24-b10; 3 414a33, 415a1-3.

87. On the meaning of *antikeimena* here ('relatives') see *dA* I 1 402b15, and Hicks 1907: 189. Cf. also *dA* II 4 416b9-11 where food is defined with reference to living beings. Cf. Hicks 1907: 339 ad 415a22: 'Nutriment in the concrete is co-ordinate as an object with sensibles and cogitables', and he adds that the 'objects' are to be studied to make the operations clear. If we translate *antikeimena* by 'relatives' it is immediately clear that their study must go with that of their relata. The question whether *erga* or parts of soul are to be studied first, is asked at I 1 402b11-15. Among the *antikeimena* of the capacities of the soul, we have the fullest account of *perceptibles* in the *PN*: *SS* 3, 4, 5; cf. *dA* III 8 431b21-432a6. The texts we have do not discuss the objects of striving (*orexis*) except insofar as these are treated in *MA* (esp. 6 700b23-701a1, 7 701a29 ff, b16 ff, 8 701b35 ff). On the connection between striving and motion see *dA* III 10. Pleasure and pain do not appear to receive a full treatment (cf. *dA* II 3 414b4-6, *SS* 1 436a9-11).

88. *dA* II 4 415a14-23. This argument seems to be in glaring contradiction with the way he in fact proceeds in *dA*.

89. What does 'prior' mean at II 4 415a18-20? Hicks (1907: 339 ad 415a19) suggests 'prior to us' (as in e.g. *Ph.* I 1). This would account for the order of exposition (note *kata ton logon*, lines 19-20). But texts such as *Metaph.* IX 3 on capacities, and above all IX 8 (referred to by Hamlyn 1993: 95 ad 415a14) on the priority of activity to capacity would suggest that there is an ontological priority as well.

90. *dA* II 4 415a26-b7: note the ambiguity in *trophê* here – 415a22 it is parallel to *noêta* and *aisthêta*, and so means food; in a23 it is parallel to reproduction (*gennêsis*), and so means the activity of nourishing; we discuss the ambiguity below; on reproduction and nutrition cf. 416b15-17, 23-5 and 416a19.

91. Cf. Hicks 1907: 339 ad 415a26. Cf. also *GA* II 1 731b24-732a1, III 10 760a35.

92. *dA* II 4 416b24-5.

93. Trendelenburg (1877: 132) compares *GA* II 3 736b26: sperm is a residue.

94. 415b8.

95. Cf. also *entelecheia* 415b15.

96. *dA* II 4 415b24-8.

97. Cf. *PA* I 1 641a27, on nature, i.e. the soul, as the substance of natural things; 641b5 on the soul as the moving cause of growth; cf. b9 on soul as nature.

98. 415a26. As is habitually pointed out (e.g. Hamlyn 1993: 95), this passage (and cf. *GA* II 1 731b31, *GC* II 10) is closely connected to Plato's *Symposium* 206e ff, esp. 207d, 208a; it is less often noted that this reference also allies Aristotle's psychology closely with Heraclitus: in the *Symposium*, existence is achieved through constant change, i.e. part comes into being, part perishes e.g. 207e3-5.

99. It is not just a general discussion of soul as Ross maintains (1961: ad 415b8-416a18).

100. *dA* II 4 415b23-8, cf. III 12 434a25-6; Ross (1961: ad *dA* III 9 432b8-12) notes this and goes on to interpret this to mean 'growth is due to the functioning of the faculty, and decay is its gradual failure'. 432b8-12 looks forward to *JSVM*, as does his reference to another account of why things breathe (II 8 421a4-6). On the soul as a cause of locomotion see *MA* 5.

101. 415b28-416a18.

102. Cf. *dC* II 6 288b15-18: 'All incapacities in animals are against nature, like old age and wasting away. For perhaps the whole constitution of animals is constituted from things that are distinguished by their proper places. For none of the parts occupies its own place.' Gill (1989b: 212) uses this passage as a main witness for her view that decay goes against the nature of the living thing and is due to its lower material properties. Before accepting Gill's conclusion, we should first note what kind of passage it is: a parenthesis in *dC*. It is thus not the right place to look for the account of the ageing of animals. In particular, we should not expect to learn anything about animal nature. Incapacity may be against the nature of the animal. But this says nothing about what initiates the process of ageing: the factor determining the composition of the animal. This is certainly not an element: there is a new nature in control, that of the animal. Nothing, however, is said in *dC* about the animal's nature and so a whole level is left out. This is, perhaps, not surprising: *dC* is not about animals.

103. *dA* II 4 416a9-10, a13-18.

104. 416a16-18; see below 4.2 on *GA* IV 10, for a discussion of the relation between size and length of life.

105. He returns to nutrition at 416a19, thus keeping his promise at 415a23.

106. Noted by Hamlyn 1993: 95.

107. 416a22-5, 29-b3. On this question cf. *Ph.* VIII 7 260a29-33.

108. *GC* I 7 esp. 323b1-15. Assimilation: 324a9-11; a common range as the basis for interaction: 323b18-29; but interaction also requires difference within the range 323b29-324a14.

109. See 416b3-9.

110. 416b9-11.

111. 416a24, 30 f. A note on Aristotle's terminology (largely following Buchheim 2001: n 16): *phyesthai* is the formation of a living thing, up to the completion of its form (*Ph.* II 193b12-18, *Meteor.* II 2 355b6-11, 3 358a17, *GA* IV 10 777b35); *auxesthai* is an increase in size of the formed thing *GC* I 5 320b30, 321b10-322a15). However, *phyesthai* can also refer to an increase in size e.g. *dA* III 12 434a26 where *phyesthai* is contrasted with *phthinein*; and is even used to define *auxesthai Metaph.* V 4 1014b20 (above §§2.1, 2.2). Both *phyesthai* and *auxesthai* (together with *phthinein* and *akmê dA* III 13 434a24) are part of *trophê* e.g. *GA* II 4 740b29-741a2 (quoted above p. 21).

112. 416b11-15.

113. 416b17-20. At 416a22 Aristotle has restricted the relevant contraries to those just providing growth, and not merely generation; all that he wants to do here is restrict growth to contraries in the category of quantity (416a25).

114. The question of the reading at line 25: *hôi trephetai* or *hôi trephei* is cleared up by bearing in mind that the subject must the same as that in the previous sentence: soul, and the soul feeds sc. the body; so we read *hôi trephei*. (This reading follows Themistius at b25-6 but he takes *hôi trephei ditton* as referring to nutritive soul and innate heat, since then *hôi trephei* will not be the same as *hôi trephetai* (lines 22-3), as Ross' reading demands.) This can then be taken to refer to the cold and hot, which Aristotle elsewhere calls instruments of nutrition (on nutritive soul using a tool (hot and cold) also *GA* II 4 740b35 (quoted above p. 21), 6 743b27, *PA* II 3 650a3, 4 651a8 ff (on blood), 7 652b16; on the hot as active *Meteor.* IV 1 379a14, *GC* II 2 329b25); natural heat burns food, but must be cooled, so that it does not go out. The second choice involves line 27: are we to read *to kinoumenon monon* or *to kinoun monon.* If we are talking about the instruments of nutrition, then it does not make much sense to say that one both changes (transitively) and is changed, whereas the other is only moved; for how can something only moved serve as an instrument? So we choose *to kinoun monon*: heat changes the food, and is cooled, whereas the cold only changes the heat.

115. Cf. *dA* I 4 408b14: a human sees *by means of* the soul (dative).

116. 416a34-b2. On the body as a tool for the soul see *dA* II 4 415b15-20. I am following Trendelenburg (1877: ad loc.) and Cassirer (1932: 61); the latter draws attention to 416a34.

117. 416b28-31.

118. *GC* I 7, *Metaph.* V 12, IX 1.

119. Cold is left out at 416b29, as throughout his treatment, the emphasis is on the likeness of the assimilation to burning. It should be noted that hot and cold are anyway relative to one another; on bodily heat as a mean, see below §4.3. Gill (1989b: 232-4) does not mention the functional role of heat in the heart. This is in keeping with her divorcing the 'active capacities' of the soul from the 'passive capacities' of the proximate matter. It is also in keeping with the rather rarefied notion of nourishment she has: 'The main function of nutritive soul is to regulate the full range of higher activities so as to maintain the balance of life' (1989b: 234). It seems simpler to say that the main function i.e. activity of nutritive soul is to nourish the living being and to reproduce it.

120. The clearest references to a work on nutrition are *SV* 3 456b5 and *PA* IV 4 678a19, but also *dA* II 4 416b31 and *PA* II 3 650b10, 7 653b13, *GA* V 4 784b2. *Meteor.* IV 3 381b13 says nutrition is talked about elsewhere. (This list of references is from Bonitz 1870: 104b16ff.) Louis (1952: 35) suggests that only a prototype was written. Most of his paper is concerned with dating of writings, using Nuyen's (1948) chronological scheme of the psychological writings. He thinks that *SV* 3 refers to *PA.* Nussbaum (1978: 375-6 ad *MA* 10 703a9-10) tends to think no such treatise was written. Only two of the references to a work on nutrition suggest at all clearly that there should be a *separate* work on it (*SV* 3 456b5 and *GA* V 4 784b2). But even such references, of the form *en tois peri* X do not imply a separate work entitled *peri* X. See Rodier 1900 vol. II: 247 for evidence. One central topic for such a treatment would be the preservation of pneuma, along with the other parts of the body (cf. *MA* 10 703a15); see below §4.7. Another would be the relation between nutrition and the stages in life (*GA* V 4 refers to a work, if it is a work, 'on growth and nutrition': see the discussion of the relation between growth and nutrition in the main text, and below n. 134 on *GA* II 6-7). There are also discussions of aspects of nutrition in *PA* II 3, IV 4.

121. See e.g. *PA* II 7 652b7-16, cf. 653b5-6; *GA* I 18 724b25; II 4 740a25 ff, esp. 740b30-41a3; III 2 752b19, 11 762b12, 763a11; IV 8 776a28, 777a22; cf. also Peck 1953: Introduction §61 ff.

122. *SV* 1 454b32 ff, *EN* I 13 1102b4 f, *EE* II 1 1219b22 f. The statement that sleep is situated somewhere between being and non-being, since being above all applies to waking because of perception, is balanced by the remark that before it comes to be able to perceive in the course of its development, an animal lives a plant's life (*GA* V 1 778b30 ff, 35 ff). Thus a sleeping animal is not simply in a state of unrealised potential (contrast Freudenthal 1995: 61, Sprague 1977): for changes that take place all the time and serve nutrition, see §§4.6 and 4.7 on breathing and heartbeat.

123. *GC* I 5 321a18-22 cf. 321b11-16.

124. *ditton* 321b20.

125. 321b24-8. These lines are important for the view that Aristotle thought matter is in constant flux, insofar as there is something changing it; see also *LBV* 3 465b25 f, *JSVM* 5 470a2-5, *MR* 1 450b6-7; *Ph.* V 4 228a9, *GC* II 10 336a16, cf. Joachim 1926: 129 ad 321b-322a4, Nussbaum 1978: 68; Frede/Patzig 1988: I 45; Anscombe 1979. In *Ph.* VIII 3 253b6-254a1 Aristotle argues that not everything is always changing – apart from anything else this would go against what is obviously the case, and changes come to an end (b25-9). For flux and life, see Wiggins 1980: vii ch. 2 esp. 85-6, 88-90. Alexander in his treatment of *GC* I 5 (Todd 1976: *de Mixtione* XVI) thinks that living matter is in constant flux (235.22, 30, 236.14-34). There is a difficulty in how to conceive of the measure measuring the water: Verdenius and Waszink (1946: 28 f) suggest a scoop. As Joachim says (1926: 130), it illustrates the flux of the flesh qua matter, and its persistence qua form. The same measure is imposed on different matter.

126. *GC* I 5 321b17-18.

127. *GC* I 5 322a5-10.

128. The faculty to make the living thing grow (*auxêtikon*) is mentioned at 322a10-16, and compared to a fire taking hold of fuel.

129. *GC* I 5 322a16-28, cf. *dA* II 4 416b17-20.

130. 320b30-3.

131. 322a20-3.

132. 322a24-8. Cf. *dA* II 4 416b11-13.

133. 322a14-16. This comparison gives rise to problems, when one considers the wasting away in the latter part of the life-cycle; if a fire is fed, it does not waste away.

134. There is no suggestion of any actual quantification of nutrition by Aristotle. Cf. also Freudenthal's (1995: 187) difficulties in accommodating the changes in quantity in nutrition – more and less; such an actuality is an all or nothing affair, as he notes, but that does not make it impossible for it to occur in differing degrees. Either nutrition happens or it does not; but more or less stuff can be involved. On growth and quantity, see *GC* I 5 322a16-28. The final lines of *GC* I 5 offer an explanation of decay, based on the failure of the capacity to nourish (322a28-33). The lines are very difficult and I can offer no satisfactory interpretation of them. I follow Joachim (1922: ad 322a28) in identifying the form in matter with the *psychê auxêtikê*. I also follow him in his emendations of *aülos* to *aulos* in these lines. I take the 'channel' thus indicated to be an example of an anhomoiomerous part, as Joachim appears to do (he refers to Bonitz 1870: 122a26 ff for the use of *aulos* to apply to organs; cf. also Alexander, *de Mixtione* 237.2, and Todd 1976 ad loc.). The comparison of the failure of the power of growth with the increasing dilution of wine by water is inexact, as Joachim says. In the case of growth failing the form remains, but in the case of the wine, it is lost: the wine becomes water (321a32). In the discussion of the formation of living things from nutrition in *GA* II 7, Aristotle makes a distinction between food involved in nutrition and that in growth (744b30 ff); both are products of concoction derived from natural food, and the food for growth is associated with the 'less pure' residues. The finest concocted stuff goes to make up especially the perceptive parts (744b12, 22), and the leftovers from this concocting go to make things of a certain bulk, such as horns and hair (744b36 ff). The point seems to be that some parts of animals do not possess sensation, and the distinction between the grades of nutrition is made with reference to this function, unlike that in *GC* (cf. Althoff 1997).

135. Living things always nourish themselves; but this does not mean that nutrition exhausts function. Contrast Nussbaum 1978: Interpretive Essay I esp. 76 ff with n 19. Cf. Sorabji 1964 on luxury functions, and Woodfield's criticisms of identifying function with self-maintenance (1976: 113-20): functions can be imagined that do not serve any useful purpose. For Aristotle what a function is, is independent of any particular function: the being of 'function' is not this or that function.

136. Cf. *PA* II 7 652b10, *MA* 7 701b13-16, *dA* III 1 425a6.

137. This is paralleled by the passage at *dA* III 10 433b13-20, referring to *MA*: there he refers to the tools of motion, and says he will talk about them 'among those functions that are common to body and soul.' This is a further indication that *MA*'s true place is with *PN* (see above n 21). Note, however, the discussion of the organs of perception in *dA* II 7-11.

138. *Metaph.* VII 8 1033a32-b19, cf. 15 1039b26-30, where the qualification is added that forms can come about even though they are not subject to coming to be; so too VIII 3 1043b14-23; cf. further 5 1044b23-4. In *Metaph.* he speaks of forms in any category as not subject to generation (VII 9 1034b7-16): thus all termini of change are not subject to coming to be (cf. *Ph.* V 1 224b4-13). An important series of arguments in *dA* I (4 408a30-b29) tries to prove that soul does not change. A further fundamental indication of the importance of change for the distinction between form and matter lies in the fact that form and matter (along with privation) are arrived at as physical principles through an analysis of change (esp. *Ph.* I 7). Change (*kinêsis*) is to be distinguished from perishing/generation. Perishing and generation do not run between contraries (*Ph.* V 1 225a32 f) – a correction of Plato's argument from cycles for the immortality of the soul in the *Phaedo* (70a4-72e2), which in effect assimilates generation and perishing to the acquisition and loss of properties, and so ensures that there is no real generation. We have here, however, the basis of the Aristotelian idea: termini of change are not subject to the change they terminate or begin – rather the changing thing changes from one terminus to the other (103b referring to 70e1 ff) Despite the distinction between change and coming to be and passing away the texts cited make clear that soul is subject to neither kind of process, as a process.

139. VII 8 1033b12-13; of course, matter is divisible, namely into portions of matter. The indivisibility meant here is that into form and matter, both of which are needed for the analysis of any coming to be.

140. See especially *Metaph.* VIII 3 1043b32 ff. This is an idea which Freudenthal has (1995: 187); but he draws the unfortunate, and as we have seen (see above §3.3), unnecessary conclusion, that development of the way in which nutrition is performed is impossible.

141. See the next section (§3.5) for use of vivisection to determine the most important organ for nutrition.

142. There is a problem about the so-called two prefaces to *LBV* (464b19-30,

30-465a2). Michael of Ephesus knew both and considered them to serve different functions in introducing the work (Michael 1903: 87.13 ff, 88.8 ff). Ross considers them to be incompatible alternative introductions. But in my view (following Siwek 1963: 270) they can be taken as compatible: the first is needed to set out the problems; the second one puts *LBV* in its place in the *PN*, although without any reasoning. 'As said before' (465a2) in the second refers back to the first.

143. On Aristotle's method, cf. Owen 1961.

144. 464b22-5. He also asks two questions about health: Is long life connected to health in naturally constituted things? (464b26-30) Are they related in the same way in all cases? He also asks generally about the place of health and sickness in physics (i.e. as opposed to medicine). Cf. *SS* 1 436a19-b1: 'Hence most of those [investigating] about nature finish with medicine and those of the doctors pursuing their skill more philosophically begin from those [investigating] about nature.' The important thing to notice here is that health can be treated in two ways: either as something to be restored, by the doctor, or as an aspect of a life form to be studied with no practical implications at all. Michael (1903: 87.6-7) emphasises that dealing with health and sickness, youth and life (this is his list) completes Aristotle's physical enquiry. The connection between longevity and health is considered in *LBV* 5 and in the *JSVM*, in the consideration of natural and violent death. See below Chapter 5.

145. 465a3-4: kinds: *hola pros hola genê* and different individuals within the species: *tôn hyph' hen eidos hetera pros hetera*.

146. 465a4-5. Following Ross 1955a: 285 ad loc. G.R.T. Ross (1908) translates 465a2 ff: 'We find this distinction affecting not only entire genera opposed as wholes to one another but applying also to contrasted sets of individuals within the same species.' He remarks that he has to translate like this, to avoid *eidos* and *genos* being used interchangeably (cf. Althoff 1992: 147). But W.D. Ross (1955a: ad loc.) points to Bonitz 1870: 151b54-6 for evidence that the two terms are used interchangeably. (This view is argued for *in extenso* by Pellegrin 1982.) Michael thinks that *eidos* is used to refer to individuals in the same species: the contrast is between horses and men on the one hand, and Nestor and Achilles on the other (1903: 88.11-15). As I use it, the word 'species' is not the modern technical term, but merely a Latin version of *eidos*. We can admit the closeness of the terms here, where Aristotle is dealing with a characteristic that subdivides species into classes of individuals.

147. Cf. Comfort (1979: 41): Senescence cannot be explained as the sum of 'environmental insult' since there is a specific age in many organisms 'which displays little environmental but marked interrace and interspecific variation'.

148. See below §4.2.

149. This is in fact the way in which Michael reads *LBV* 2 (Michael 1903: 88.18 ff, 26-31; also 87.7-10). He understands the question at 465a13 *ti to euphtharton en tois physei synestôsi kai ti to ouk euphtharton?* to mean, 'which *among* (*en*) animals perish easily and which do not?' (1903: 88.18-20), rather than 'what is it *in* animals that perishes easily?' He is alone in his reading of this passage, against Hett 1936, Ross 1955a, and G.R.T. Ross 1908, who understand it to mean: what it is in natural beings that makes them suitable for perishing or not (cf. also Le Blond 1938: 98, Tricot 1951 and Siwek 1963 ad loc.). Michael's reading gets some support from *LBV* 5 466b4 where *euphtharton* refers to a characteristic of the whole living thing.

150. Only fire and water are mentioned here, but there can be no reasonable doubt that all elements are meant cf. *GC* II 8 334b31 f: all mixed bodies 'about the centre' must include all elements. The reason that he only mentions fire and water here is that he is already thinking of the natural heat and moisture of animals in ch. 5 466a18 f, which will turn out to be the causes of their longevity.

151. 465a15-19 'reasonable': *eulogos*. Le Blond (1938: 98) classes the use of *eulogos* ('rationel') here (456b18) not as used of something explained by a theory, but with those uses marking the absence of certainty. But the rest of the chapter gives a reason for the elements passing on their character to mixtures made of them. This account can be completed from *GC* I 10 (cf. also *GC* II 3); see below §4.1.

152. 465a17-18. They do not have the same capacity. Siwek (1963: 270 n 8) says that 'capacity' here means 'property' or 'quality' – as it were that water is cold and fire, hot; but 'capacity' is a more helpful expression since it can, and here does, carry the connotation of a capacity for change sc. in another: the elements have the capacity to destroy one another, and these capacities are their qualities hot, cold etc.

153. See above p. 51.

154. 465a17-18: *metechein* 'share'.

155. On *syngenes* 'akin' see Bonitz 1870: 707b10. Here, it means that two things belong to the same genus, although specifically different – fire and dogs are both hot. One might object that if the things akin to fire and water are not earth and air (cf. *Meteor.* I 2 339a28), then these elements are not mentioned at all; and they do destroy one another. But we will see that it is warmth and wetness in particular that have to be preserved to keep things alive (see below §4.3). On contraries in living things, see §4.1.

156. The nutritional chain outside living things is sketched by Gill 1989b: 128-9.

157. Gill 1989a: 203-4, cf. 1989b: ch. 7, esp. 212 f.

158. Both characterisations occur at *GC* II 3 330b30-31a6.

159. The best explanation is the closest one (*Ph.* II 3 195b21 ff, *Metaph.* VIII 4 1044b1). This touches the question of compounds, and their relation to Aristotle's mixtures (see below §4.1).

160. On *synthesis*, see *GC* I 10 328a6 ff. Siwek 1963: 270-1 n 10 refers to *Ph.* I 7 190b8 for the sense of synthesis as the juxtaposition of parts, and distinguishes it from *mixis* and *krasis*. On the living body and *mixis*, see below §4.1.

161. 465a18.

162. 465a13 f, 17 f, cf. *physika* 24, and 27: 'if the soul were not *by nature*, but just as knowledge in the soul, so the soul in body ...'

163. 465a22 *dektikôn*.

164. 465a23-6.

165. 465a27-32.

166. Then they are no longer truly parts of the body, except homonymously: the corpse is only homonymously a human and hence not mentioned in *LBV* 2 (*Metaph.* VII 10 1035b24-5, *Meteor.* IV 12 389b30 f; cf. Gill 1989b: 129-30; 162-3).

167. At 465a27 whether one reads *logisait'* (Siwek 1963 and Ross 1955a with most Mss) or *syllogisait'* (with G^{a1}, followed by Bekker and Biehl), it seems hard to get more than an informal argument from the text (i.e. not a formal syllogism).

168. See esp. Freudenthal 1995: 191.

169. Contrast Freudenthal's remark (1995: 13) about *LBV* 2 465a12: 'Aristotle construes destruction as the inevitable outcome of material necessities'. This might mean Freudenthal thinks that soul is being excluded from consideration.

170. *koinônia* 465a30-2. Is it in reference to the functions shared by (*koina*) body and soul (see above §3.1 p. 35 f) that Aristotle talks of *koinônia* here? *dA* I 3 407b18 ascribes things that living beings do and undergo to the partnership between body and soul, in the context of criticising predecessors who gave no account why the soul is in body – and we have seen how closely this line of thought is connected to the project of *PN* and *JSVM* in particular (see above §3.1). Aristotle, of course, does not think that soul is an epiphenomenon of body (cf. Ross' remark (1955a: 286) on 465a27-8), rather that body must take part in vital activity. In the *Eudemus* (Frg. 7), one reason that soul is not a harmony is that there is a contrary to harmony, but not to soul. This fact, of course, only accentuates the need for the connection between soul and contraries to be made clear, if things with soul (living beings) are to come to be and perish, and all coming to be and perishing involves contraries. Another aspect of this partnership will be discussed in Chapter 5 on coming to be as one of the stages in life. Cf. Michael of Ephesus' (1903: 89.15-29) account of the end of the chapter: if the soul were not the form of the body, i.e. if it did not come with the seed but came later, that is to say when the living things already exists, there would be a perishing proper to the soul. He refers to the *dA* (e.g. II 1 412a20) for Aristotle's view that the soul is the form of the body.

171. See above §3.1 p. 40. For a later Peripatetic discussion of the seat of the soul, cf. Alexander *de Anima* (Bruns 1887) 94-100.

172. See above §2.3.

173. As Ross (1955a: 295 ad 467b13-15) says, the reference is to *dA* II 2 414a19: 'it is a good supposition that the soul is neither body nor without body. It is not a body but something of a body, and because of this in a body'. Other references to *dA* II in *PN* are at *SS* 436b10, 14, 437a18, 438b2, 439a8, 16, 18, 440b28, *SV* 454a11, 455a24, *JSVM* 14 474b11 (on this chapter see p. 98). Ross (1955a: 17) suggests they are later additions; but as is clear, the back reference in *JSVM* 1 and that in ch. 14 fit well with a unitary view of *dA* and *PN*. It would be hard to motivate an enquiry into the seat of the soul with no previous treatment of the soul, and there is no reason to think the one referred to is not the one we possess.

174. On this sense of 'in' cf. *Ph.* IV 3 210a20-1 'as health is in hot things and cold things and generally the form in the matter', and the corresponding sense of 'have' ('the matter has this form') *Metaph.* V 23 1023a11-13.

175. Cf. Alexander's comment (1901: 5.13-17) on how the things common to body and soul are to be investigated in *SS*: by looking at the parts of the body 'in which and through which the activities (*energeiai*) of things with soul [exist]', and he refers to *dA* for the necessity of body to soul (see above §3.1 p. 40).

176. On this sense of *dynamis* see *Metaph.* V 12 1019a15-22, IX 1 1046a9-29. This is clearly connected to the concept of an *archê* (*Metaph.* V 1 1012b34 f, 1013a4-6). One must distinguish here between the part of the body from which the change issues (e.g. 467b34 f, 468b22 ff), and the soul as the principle from which change in the whole body originates (*dA* II 4: above §3.3, and cf. §2.2). The soul is often called the nutritive or sensitive principle, and is in a part with a principal role in the relevant actuality. This is the role of the common sense organ 467b28, 469a10-23: all the sense organs can cause change in it; the principle of the nutritive soul is said to be in the middle part of the body (468a21); so too 469a5-8: the principle of both souls is in the heart, and 469a25-8; at 468b4 'perceptive principle' is used as the parallel to 'nutritive soul' at 468b2; the perceptive soul is a principle in the heart at 469b5-6.

177. The part that takes in food, and that which produces excrement can function best if the principle is in between them (469a29-33); on the connection of perceptive parts to the common sense organ see 469a10-20.

178. Lines 467b17-19 dismiss the other faculties or parts of the soul – i.e. especially reason, which is not relevant to the discussion of the faculty in virtue of which animals are said to live. The faculty of locomotion and its parts are mentioned in the schematic division of the body in 2 468a18-20.

179. 467b18-27; for parallel formulations about things that live and animals, see *IA* 4 705b8, *SS* 1 436a3, *PA* IV 5 681a13, *dA* II 2 413b4.

180. *einai*. Michael 1903: 100.14 says they are different in account. Presumably this follows from their being different in being. For the identification of being primarily with actuality cf. *dA* II 1 412b8-9.

181. Ross 1955a: 295 ad 467b18-22.

182. There is a parallel to this argument in *JSVM* 3 469a17-20, where a very similar argument is used to the one at 467b18-27: if the centre of nutrition is in the heart, then that of perception must be too, since the being lives insofar as it is an animal, and is an animal insofar as it perceives.

183. *dA* II 3 414b20-415a13.

184. Michael 1903: 100.29-101.7. Cf. *dA* III 13 435b4.

185. See above §2.3 p. 33.

186. Cf. *MA* 10 703a29-b2.

187. Cf. Michael's introduction (1903: 100 esp. 16-29), in which he concentrates on the heart to the exclusion of living things without hearts.

188. Cf. *archê MA* 10 703a37-b2.

189. *PA* II 13 657b18-21, III 4 665b14-16, 666a14-17.

190. On the dimensions of living bodies see esp. *IA* 4, 5, 6, *PA* II 10 655b29-37, and in general Lloyd 1992: 154.

191. Cf. *PA* II 10 655b30-37 on the middle part of animals being the third part. On the middle see Lloyd 1966: 53, 64, Tracy 1969: 184, Byl 1980: 238-50. For locomotion in relation to forwards and backwards see *IA* 4 705b11, *PA* III 3 665a14; on the oblique motion of crabs, and their vision, see *IA* 14 712b13-21, 16 713b12-17.

192. 467b28-30. See Lloyd 1978: 233-8 for a critique of the indications Aristotle used.

193. 468a1, cf. *IA* 4 705a29-b1.

194. This distinction shows that Aristotle is concerned both to link living things to their surroundings, but also to assert their independence: living things can be the same way up as the cosmos (that is part of the superiority of bipeds, *IA* 5 706b2-13), or not (plants). At one point he distinguishes explicitly between an orientation by position, i.e. according to the surroundings, and one according to function (*IA* 4 705a30, b17) which is what we have in the opening chapters of *JSVM*. Cf. *Ph.* III 5 205b31-4.

195. On plants in comparison with animals see §4.3 on *LBV* 6 and §4.5 on *JSVM* 6.

196. On the upright stature of humans, making them most correspond to the upwards direction of the whole (468a5-8), cf. below (§4.6) on humans' greater heat, which is of course (!) responsible for their uprightness.

197. *teleia* 468a13, see *Metaph.* V 16 esp. 1021b21-5; completeness or maturity is closely connected to the natural capacity to reproduce (*dA* II 4 415a27, III 9 432b21-5).

198. 468a13-20.

199. As Lloyd remarks (1975: 179, 190), vivisection is embedded in a research project into vital functions. Ross (1955a: 297 ad 468b12-15) says that only here, *JSVM* 479a3-7, and *HA* 503b23-7 are there clear references to vivisection in Aristotle.

200. Following Ross (1955a: 301 ad 469a23-b1) against Ogle.

201. 468a31 f.

202. 468b11-14.

203. Aristotle calls the results of the primitive experiments involved here 'phenomena' (*phainomena*, 3 469a24), obviously meant to correspond to perception, as opposed to the general argument (*logos*) mentioned already *JSVM* 2 468a20-3, which is then discussed in Chapter 4 below, p. 95; cf. Michael 1903: 101.18f. On the varieties of *phainomena* in Aristotle see Owen 1961: esp. 115; Lloyd 1979: 136-8.

204. See below §4.3 on *LBV* 6 467a22-6 – the capacity of plants to produce stalk and root from any part is made responsible for their longevity: they can rejuvenate themselves in this manner.

205. Aristotle's cardiocentrismus was based in great part on the arguments we find in *JSVM* and *PA* III 4, *HA* VI 3. The reference at *JSVM* 3 469a20-3 for the reasons that some sense organs are in the head is to *PA* II 10 656a27-657a12. Alcmaion's early rejection of the heart in favour of the brain (cf. Theophrastus *de Sensibus* 4.26) had found favour with a distinguished set of Aristotle's predecessors – Plato (*Timaeus* 73b1-e1), Democritus (DK A105, C6 but cf. B1), Diogenes of Apollonia (DK A19, 21), and Hippocrates (according to Galen *de usu partium* I 8).

206. As Michael remarks 1903: 103.9-10; cf. the lower animals that are like many animals put together 468b10.

207. Cf. *GA* III 2 752a18-23.

208. Ross (1955a: 299 ad 468b18-24) cites opinions from the nineteenth century which speak of the point where root and stem join as the plant's 'heart'.

209. Of course, the heart in humans is not in the centre, but to the left of centre; see Lloyd 1978: 244; for Aristotle's explanation see *PA* III 4 665b20-7.

210. Tracy 1983: 332

211. 469a5-7. On the heart as the primary sensitive part cf. *PA* III 4 666a34 ff, *GA* II 6 743b25 ff.

212. On sanguineous animals, and man in particular, as models cf. sp. *SV* 2 455b31 ff; cf. Lloyd 1983: I.3 'Man as model in Aristotle's zoology'.

213. 468b29 f. *PA* III 4 665a33-b2, *HA* VI 3 561a11-13.

214. *GA* II 5 740a3-13. Cf. Tracy 1983: 333-4.

215. Peck (Appendix B to his 1953) claims this is Aristotle's theory. See Lloyd 1978: 236-9 on difficulties with Aristotle's account of the transmission of sensation. For a detailed, if inconclusive discussion of the channels to the eye see Johansen 1998: I 10. For pneumatised blood as a transmitter of sensation cf. below §4.7 n 283.

216. Neuhäuser 1878: 120 ff.

217. *JSVM* 3 469a14. Contrast *PA* II 3 650a18: there must be channels from the stomach to the rest of the body; these are then identified with blood vessels on the basis of dissections (a29).

218. Ogle 1897: 66 on 469a9-10.

219. Michael 1903: 105.16-20. On the parts involved in working on food, see *PA* I 3 650a3-650b13; *GA* II 6-7; on spleen and liver cf. *PA* III 7 670a20, 33.

220. The heart is the locality where sensations are had and that without which no sensation at all would be had. The final part of chapter 3 (469a10-23) offers arguments for the placing of the master part of sensations in the heart: touch and taste have their central organ in the heart and thus, by an argument from economy, all the senses must (following Ross 1955a: 299 f ad 469a10-16). There is a subsidiary argument (469a14-16), which is very obscure: the others senses can cause motion in the centre, but touch and taste do not extend to the 'top' i.e. head. On the senses in the head and heart there is a reference probably to *PA* II 10 656a25-657a12 at *JSVM* 3 469a20-4. On the sense organs in the head cf. also Michael 1903: 106.28-9 with his reference to *GA* (II 6 744a2 ff). At 468b32-469a1 the reference is probably to *PA* III 4 665b15-16, at 469a1-2 to II 4 651a13-15.

221. Tracy 1983: 332-3

222. The interpretation of *JSVM* 4 continues below in §4.4 p. 95.

4. Body

1. Aristotle argues against identifying the soul with *harmonia*, an idea connected to that of a balance (esp. *dA* I 4 cf. also Frg. 7). This does not mean, however, that soul and *harmonia* are unrelated. For while the soul is not itself a harmony, it is thought to be found in a 'harmonious' body, that is to say, one that is fitted together in a certain way (which is what *harmonia* means). Cf. on harmony and the mean Plato *Philebus* 25a ff esp. 26b; against the soul being a *harmonia*: *Phaedo* 93b-d. On physiology and the mean in Aristotle, which is also closely connected to balance and *harmonia*, see Tracy 1969: IV A, B, C; in connection with a balance between basic qualities, Althoff 1992: 71, 88, 113, 164. An important argument used by both Plato and Aristotle against the soul being a *harmonia* is that it then could not cause changes; this defect does not apply to Aristotle's idea of an activity.

2. Cf. the treatments in Gill 1989b: 147 f, Bogen 1995, Fine 1995, Code 1995.

3. *PA* II 1 646a12-24, cf. *GA* I 1 715a14 ff.

4. *PA* II 1 646b6.

5. Above, especially in §3.2, *energeia* was translated by 'activity', *entelecheia* by 'actuality'; here it seems better English to say that in a mixture containing fire, the fire is not actual fire, although the heat of the fire is active, i.e. as a capacity. So I have decided to say that the ingredients in a mixture are not 'in actuality', instead of 'in activity' for *energeiai*.

6. *GC* I 10 327b29-31; see Bogen 1995: 389, and Fine's criticism 1995: 286-7.

7. When the capacities for change are near enough to work on one another, they do so (*Metaph.* IX 5 1048a5 ff). In *GC* II 4 the transformation of the elements into one another is described. In any transformation between elements that only differ in one quality (from fire – the dry-hot – to air – hot-wet – by wetting), only this one quality changes. But the resultant element is still seen as dominantly characterised by one quality (cf. II 3 331a3 ff).

8. Cf. Joachim 1926: 178 ad 327b10-13.

9. Usually (e.g. Waterlow 1982: 85, Gill 1989b: 146-7), the preservation of capacity in mixtures is taken to mean above all that the ingredients can be recovered from the mixture (because of the passage 327b27-9): they are still there in capacity, because they can be recovered. This aspect is not important for the question of the capacities of the mixture, which are at stake here. (Both are aspects of the idea that in one respect nothing is annihilated by change.)

10. *GC* I 10 327b33-328a18.

11. *GC* I 10 328a28-31. The kind of things that mix well are those that are easily divided, since when small parts of the ingredients are side by side a mixture comes about most easily. These are primarily liquid (328b3).

12. Esp. 334b10-16, following Joachim's interpretation (1904: 76). There is no need to exaggerate the comparison with a modern chemical formula, as Joachim does (see Wardy 1990: 258); nor should we think of expressing as a number the proportion (*logos*) between the capacities present in a mixture. Aristotle nowhere suggests such a move; for an optimistic view of measurement in Aristotle's *Physics*, cf. Hussey 1983: Additional note B 'Aristotle's Dynamics', 1991: esp. 222.

13. On action and passion, see *GC* I 7.

14. *GC* II 7 is devoted to the question of how uniform parts of living bodies arise from the elements; for the idea of balance between the basic capacities, see esp. 334b8-30; Aristotle speaks of a middle (334b13, 27), and a mean (29); cf. also 23, where he says that if the hot and the cold do not balance one another (*isazei*), one will just be destroyed.

15. The contraries in mixtures are relative to one another (334b9 f, 15-16). The definition of homoiomers through the ratio of the ingredients is Joachim's idea (1904: 76). It does not banish the problems with the localisation of the different capacities in a mixture; but it does emphasise that there is an answer to what constitutes the mixture on the basis of the ingredients.

16. Cf. Joachim 1926: ad 334b14-16.

17. Cf. Joachim 1926: 244 ad 334b28: 'For this familiar conception of a *meson* which is capable of fluctuation with in certain defined limits, cf. *EN* e.g. 1106a 26-32, 1106b36-1107a2, 1173a23-8.'

18. Waterlow 1982: 83-6.

19. Waterlow 1982: 86.

20. Cf. Waterlow 1982: 85: 'The independent natures of the four elements are not totally "suspended" in flesh, bone etc., but are permitted limited expression, the bounds of which are set by organic functioning.'

21. Waterlow 1982: 86, summarised above in the main text. Wardy 1990: 256. His ch. 6 is devoted to the problem of reduction in Aristotle.

22. Among things that exist, elements and parts of the body qualify as capacities, rather than substances (*Metaph.* VII 16), i.e. independent actualities. Thus elements and parts of bodies can remain what they are, capacities, within a greater whole. *Metaph.* IX 1-9 is devoted to capacities for change and the way such capacities are involved in actualities and activities.

23. For the sake of the analogy I ignore the condition that the ingredients of a mixture must be separated out on dissolution of the mixture (*GC* I 10 327b27-8).

24. The chapter begins by asking about the destructibility of fire in the upper regions (465b2), that is, where it is not confronted with a contrary. This apparent cosmological slant of the chapter is rather misleading: Michael of Ephesus thinks chapter 3 is about the aether. According to him, the question is whether it can, although naturally perishable, be made imperishable by its place (1903: 90.7, 91.18-19). I think the cosmological side of the problem is rather the topic of *GC* II 10.

25. My reading follows Michael 1903: 90.19 ff.

26. Ross (1955a: 288 ad 465b1-25) translates the phrase by 'the contraries *among* substances' rather than the contraries *in* substances. Michael thinks that substances here are 'animals and all composites and in them the contraries [are] the four elements. Therefore, since everything that is destroyed accidentally is one of those things in a subject, no element would be destroyed accidentally' (Michael 1903: 91.6-10). He finds the reason for this in the equivalence of substance *being predicated of* no substrate with its *being in* no substrate. This last point raises great problems in relation to the *Categories* (2 1a20-b3), where being predicated of no substrate and being in no substrate are indeed characteristics of substances, but distinct ones.

27. 465b10-32. Michael (1903: 91.20) takes the second half of the chapter to be the solution to the problem in the first half. This is clearly better than Ross (1955a: 288), who thinks that from b10 'Aristotle proceeds on the opposite tack', i.e. claims that all things with matter after all do perish. Nonetheless, Ross thinks that the conclusion of the second part has been proven. Yet it cannot be, as long as the conclusion of the first part stands, since it leaves the possibility of non-perishing open.

28. There are two versions of this: it is impossible for something with matter not to have contraries in some way (465b11-12); and matter has contraries immediately (b29-30). The second statement is the more fundamental: because matter has contraries immediately, anything with matter has contraries.

29. 465b4-5.

30. 465b12; Ross (1955a: ad loc.) emends *pantei* to *panti*. The original reading should be kept (following Michael 1903: 92.18 and Siwek 1963).

31. Michael 1903: 92.26-7, 92.30, 93.1, 10, 12. Cf. *GC* I 7 324b18-21: perhaps heat must be in matter, that is to say with heat is always given the possibility of its loss by the thing. If the heat has to be in the stuff for it to be that stuff, then the loss of the heat will mean the destruction of the matter.

32. Michael 1903: 92.18-93.23. Similarly Siwek (1963: 272 n 14 on 465b10 ff).

33. Michael thinks that matter 'fights form', that is to say, tends to push off one contrary and so accepts another (1903: 91.28-9). This is a Platonic view of matter (cf. *Timaeus* 49d-e, 50c). Freudenthal seems to share Michael's view when he speaks of the 'inherent instability of the equilibrium of powers within all composite substances' (1995: 13).While Freudenthal mentions the accelerating or braking effect of the environment on perishing, he does not say whether he envisages a moving cause as necessary for change.

34. 465b14-17. For the account of active and passive capacities, and their action on one another when they get close to one another cf. *GC* I 7, *Metaph.* IX 1 1046a9-19, 5 1047b35-1048a7. The necessity here is simple insofar as an unvarying connection is asserted (*Metaph.* V 5,1015a33-b9, VI 1026b27). Simple necessity applies to that which is invariable and changeless (1015b14 ff). See Cooper 1985: 165 esp. n 12. For a denial that causes (such as the capacities for change here) necessitate see Sorabji 1980 ch. 9, and also pp. 52-3 on our passage; Aristotle's specifications allow for no more regularity than what usually happens. The point is of course that while Aristotle in principle has a strict notion of necessity (whenever x, then always y), he cannot, and knows that he cannot, specify the conditions, and so allows for latitude (*Ph.* II 7 198b6, *APr.* I 13 32b6-22). Hintikka 1973: 202 takes our passage to say that natural tendencies have to be prevented from realising themselves (with *Ph.* VIII 1 251a23-8, 251b5-10, *MA* 4 699b29). The attractive aspect of this thesis is that it would give an answer to the question of why preservation is needed (cf. on flux above p. 160 n 125); but it would seem to make Aristotle's capacities for change unnecessarily mysterious. For a statement of what is needed for capacities to be realised, see *Metaph.* IX 5.

35. *perittôma* is the remains of food (*GA* I 18 724b26). Nothing is fed that does not participate in life (*dA* II 4 416b9). See Bonitz 1870: s.v.; Bonitz (1870: 585b43) links this use at 465b17 with *PA* IV 2 677a27, and *GA* II 4 740b8. Cf. Siwek (1963: 272 ad 465b17) who sees it as meaning generally 'residue', and thinks that fire is not imperishable since it too produces residue, that is, ash.

36. 465b19-21.

37. 465b22-3. We should perhaps note that destruction is not merely a side-effect of otherwise beneficial matter: the end or good is not mentioned in this passage. One might take it to be implicit in the talk of residues, i.e. what serves no good. But the point is a general one about contraries and process. Cf. Siwek 1963: 272 n 17 on 465b19-20: suppose something to have so destroyed its contrary as to produce no excretion. Is the thing then immortal? No, because it can be destroyed by its environment. Siwek and Ross (1955a: ad 465a25-6) remark that *kinêsis* here must have a wide sense 'process' since coming to be and passing away are not in the strict sense *kinêsis* (*Ph.* V 1 225a26, 32, 2 225b10-226a26). Cf. also the processes between pairs of ends in different categories at 465b31 (cf. *Ph.* III 1), and the use of *metaballein* 465b17.

38. On this necessity see below p. 114 f.

39. Ross (1955a: ad loc.) thinks that the mention of the greater fire consuming the lesser one at 456b23-5 has slipped in as a 'gloss introduced by a scribe who had in mind the similar words in 466b30-1'. But he does not explain why the scribe introduced them here. Quite how these lines fit into the rest of the passage, if at all, remains obscure. The 'hence' (*dio*) at line 23 does not seem to work. The mention of speed here is important: after all we are dealing with the relative rates of ageing of different kinds of animal. The consumption of a lesser fire by a greater one, I think, may simply be important as a case of destruction which does not seem to fit the pattern of one contrary being destroyed by another. So Aristotle has to explain it by the exhaustion of fuel. This would fit with Michael's similar idea (1903: 94.10 ff): the lines prove that fire feeds and hence produces a contrary, and hence perishes. Cf. also *dA* II 4, 416a10 ff, 416a27.

40. Cf. Michael 1903: 91.6-10. Michael takes the expulsion also mentioned in the chapter to be about the expulsion by fire of its waste and not that by a living body (*exelaunei* 465b19, Michael 1903: 93.24-94.9). This is, according to Michael, Aristotle's answer to an objector who claims that fire has a separative faculty (*apokritikê dynamis*) like animals, i.e. a capacity to produce waste, and thus to separate itself from its contrary and so remain indestructible. Aristotle's answer is then, according to Michael, that fire is either destroyed by air, that is, by its environment, or that there is some actual contrary, i.e. residue, still remaining within it which then destroys it. Hence perishing remains necessary for fire, as required by the purpose of the chapter. There is no exception to material things perishing. This would tell even more against reading a reductionist slant into the chapter. For we are concerned with material things quite generally, not with the relation of one sort of material things (e.g. elements) to another (e.g. living things).

41. 465b27-9. Note also growth and decay at 465b31.

42. 465b11-12. This fact is cited again at the end of the chapter as the reason for the necessary decay of things (465b30).

43. 465b27-9. It is worth contrasting this treatment of change in living things with the brief treatment in *Ph.* VIII 6 esp. 259b6-12, where the emphasis is on changes in animals – growth, decay and breathing – insofar as these are caused from outside, by the surroundings and things like food that enter the living thing (in contrast to locomotion, which is properly under the animal's control).

44. 465a9-12. Similar texts appear in *Prob.* XIV 909b34-6: inhabitants of hot regions are longer lived because they keep their heat and moisture, and death is the perishing of these things. Alternatively (909b25-31), because they have a drier nature and dryness is less liable to rot, and death is 'a kind of rotting' or 'a cooling of the internal heat' through the surrounding hot air; cf. also 909b1-8: there is slow ageing in airy places. Conditions under which there is less rotting are listed at *Meteor.* IV 1 379a9-b9, e.g. cold weather 379a26-9. Alexander explains this as follows (1899: 185.5, 15 ff ad 379a19): in the cold, the environment is not hot enough to take control of the hot in the thing concerned. Cf. Freudenthal 1995: 40-2.

45. *Ph.* VIII 6-9 cf. *MA* 5.

46. On the dependence of living things on the movement of the sun, especially see *GC* II 10. Cf. *Metaph.* XII 5 1071a13-17. It is worth noting that in *LBV* Aristotle nowhere mentions the lengths of lives (in figures) of different living beings. For difficulties in doing this even now cf. Comfort 1979: ch. 1.1. The difficulties start with what exactly one is meant to be measuring. Note that Aristotle does not say that any species must (essentially) live *this* long.

47. See *GC* II 10 for cycles of elements turning into one another 337a3-6, cf. *Meteor.* I 9 346b35-347a8; on cycles in human affairs in common parlance *Ph.* IV 14 223b24. See also Preus 1990: 484-8.

48. Williams 1982: 191 ad 336b10.

49. 336b10-15, Revised Oxford Translation (Barnes 1984).

50. Williams 1982: 191-2. See the definition of a measure *Metaph.* X I 1052b18 ff. On the use of the heavenly revolutions as a temporal measure see *Ph.* IV 14 223b12 ff.

51. Cf. Joachim 1926: 261 ad 336b10-15.

52. The *locus classicus* is *Metaph.* XII 10 esp. 1075a11-15; for nature as the cause of order, which accommodates irregularities *Ph.* VIII 1 252a11-19, *GA* III 10 760a31, *MM* II 8 1206b37 ff. Natural order is otherwise commonly connected to changes (*Meteor.* I 14 351a25, *dC* III 2 300b31-301a12).

53. See Philoponus' remarks (1897b: 292.24-32); on heat destroying heat cf. above n 39.

54. Generation and corruption are here the gradual processes of growth and wasting away (Philoponus 1897b: 293.13). This makes sense since generation and perishing of complex things are not to be understood as instantaneous processes. There are unhappy aspects of the theory in *GC* II 10, such as the equal duration of coming to be and passing away (336b18 f), and the (obscure) explanation of exceptions to this rule, which we can pass over here.

55. See above §1.3 on cycles. The cyclical nature both of climatic changes and of the generation of living things are said to depend on the revolution of the heavens (*GC* II 11 338a17-b11). Williams 1982: 193 notes the connection between this text and *dA* II 4 415a25-b7, on coming to be as a way of being, and also with Plato's *Symposium* 207d ff. Cf. above §3.3 p. 50 and n 98.

56. This is one of the texts in which Aristotle speaks of God ordering things in such and such a way (esp. 336b30); cf. Balme 1987b: 278. On teleology see above §2.3.

57. *Ph.* IV 12 221a26 ff, 13 222b16 ff.

58. Hussey 1983 emends the text at 222b22-4, at 23 reading *pathein* for *prattein*, translating 'nothing comes to be without its being changed in some way and being acted upon (instead of: acting), but a thing may cease to be even though it is not changed' *scilicet* from outside, so suiting his reading of the action of time.

59. *Ph.* IV 12 221a8, cf. Hussey 1983: 166, and *dC* I 10-12, *Metaph.* X 10.

60. See below §4.6 for the connection between food and surroundings. The separation between moving cause (especially heat from the sun), and matter (especially food) is tricky in that the heat of living things is connected to both; and in both cases we should resist the temptation to think of the modern concept of energy.

61. See below §4.3 on the account offered here of longevity. Peck 1953: Appendix A attempts to give a systematic description of Aristotle's global view of the relation between time, period and cycles. We restrict ourselves merely to those aspects directly concerned with life-span; Peck does not appear to have a view on the question discussed in the main text.

62. As Peck observes (1953: 478 on 777b23), Aristotle is playing on the literal meaning of *periodos*, cycle. Apart from the aspects discussed here, there are further obscurities in Aristotle's view, firstly (777b23) of the way in which the phases of the moon are related to the sun, and secondly (b24) how the 'month' is a period determined by both sun and moon.

63. Peck 1953: 480 notes the closeness of the idea of balance here with the idea of the mean in good states (*Ph.* VII 246b4) and with the doctrine of the mean in *EN* II.

64. 778a1. cf. *Meteor.* I 2 339a21.

65. 778a5-9. Cf. *GC* II 10 336b18-24.

66. Balme 1987b: 277, with reference to *GA* IV 10, *GC* II 10: 'The movements (of the celestial bodies) cause the earth's seasons and therefore exert a general influence upon growth, but nothing more detailed.'

67. 465a2-4.

68. Michael 1903: 94.29-30.

69. Balme 1962: 192 cf. 190, Balme 1987a: 85-9, Pellegrin 1982.

70. For size in *LBV* cf. the connection between heat and growth, and heat and age is important in *LBV* 5 466b16-22, 28-33. Cf. the role of size in accounts of longevity quoted by Ross 1955a: 290; and the way such theories fare in recent times cf. Comfort 1979: 193 ff. Cf. also *dC* III 6 305a7: a lesser body is more perishable than a greater one – i.e. the rule does not seem to depend on the life of the body concerned; cf. also *JSVM* 14 474b15-19. On heat and resistance to decay see Freudenthal 1995: 40-4.

71. 466a17. Note the prominent position of cause (*aitia*) at the start of the chapter. Cf. Siwek 1963: 273.

72. 466a18-20. These lines also serve to underline the connection between *LBV* and the topics in *JSVM*.

73. See above §3.3. Althoff 1992: 148, 152 notes this connection here; cf. *PA* IV 3 677a30 where the lack of gall is said to make the animal concerned long-lived (see below and cf. *APr.* II 23, *APo.* II 17 99b4-7), for reasons connected with nutrition; conversely, very large teeth would be required if life lasted a thousand years (*GA* II 6 745a29-b2); the greyness of hair in old age 'which is cold and dry' is said to be due to failure of concoction through the proper heat of the human (*GA* V 4 esp. 784a31); see above §3.3 p. 54 f. Aristotle (*GA* IV 10 777a32-b16) also argues that size cannot be the only factor in making large and 'more perfect' animals longer lived, since humans are longer lived than anything except elephants, and yet are smaller than donkeys, and horses (grouped together as 'bushy tails'). For gestation size is of paramount importance: larger things need longer to form. This explains the longer gestation of horses. The reason for longevity Aristotle names here is the good mixture (*kekrasthai*) of living things vis à vis their surroundings (777b7 ff, cf. Plato *Phaedo* 111b1-3). He says he will speak about that later and allows that there are other natural contributory factors (*symptômata*); Peck refers in a footnote to *LBV* 4 466a15, and also to *PA* IV 2 677a35, where the liver and its specific quality in each species is said to be the decisive factor in longevity (677b1). Bile is a residue of food, is hence bitter (food is sweet), and damages the liver which is vital for animals with blood. This damage to the liver shortens the life of such animals as have bile. This passage seems to be most easily explicable by noting that Aristotle is here providing a (rather *ad hoc*) naturalistic justification for the 'old' saying that one lives longer if one has no gall (677a30). The theory in *LBV* does allow a relation between the mixture in living things and their surroundings; but this relation does not ground specific life-span, as it would seem to in *GA* IV 10, rather how long something lives within the limits of specific life span (see esp. *LBV* 3 465b27-9).

74. Contrast Siwek's translation of 466a20 (1963): 'manifeste patet hoc [ex experientia]' (Siwek's addition). A good example of an educated appearance is at *JSVM* 20 477b9: water animals are clearly not (*phainetai ... onta*) hotter than land ones, since some have no blood at all, others only a little (followed by a reference to the discussion of when one thing is hotter than another at *PA* II 2 648b2).

75. *Pace* Freudenthal 1995: 191.

76. 466a20-2, cf. *PA* II 1 646a16, *GC* II 2 329b26-32, see above §3.4.

77. 466a23-5; for fat and nutrition cf. *SS* 4 442a17, 8. On *lipara* cf. *PA* II 5 651a25 f, and their connection to *pneuma GA* II 2 735b27-30.

78. 466b2-4. cf. 467a3 f. On oil, cf. Freudenthal 1995: IV.2: 'Resisting decay: Fatty moisture and oil'; 'By virtue of their "interlocking parts", oily substances are highly cohering. They do not admit of the separation of the moist from the dry components' (p. 163). 'Interlocking' is an unfortunate choice of word since it suggests quasi-atomic parts locked together. Instead, cohesion in Aristotle's mixtures must be a question of the relation between the capacities (see above §4.1).

79. 466a13-14.

80. 466a1-2. Cf. also Freudenthal on the connection between wet and hot (1995: ch. IV.1.2 esp. 155).

81. 466a33-b2. Michael (1903: 95.30-96.1) suggests that if horses have three times as much wetness, yet man is ten times as hot, then man will be longer lived.

82. Siwek 1963: 273 n 29. Cf. *PA* II 2 648a19-649b8 on the difficulty of deciding whether one thing is hotter than another.

83. *LBV* 3 465b16-23 also talks about the destructive powers of residues, working as contraries on the living thing. For the heat of living things producing residues through concoction, see *GA* III 11 762b6-9, and cf. *LBV* 5 466b6.

84. Cf. *GA* V 4 784b32 ff, where old age is called a natural disease and disease accidental old age, insofar as some diseases produce the same effects as old age. Health and sickness were also mentioned as being related to longevity in some cases in *LBV* 1 464b26-30; health and disease are then (464b31-3) said to belong to the course of enquiry in *PN*, insofar as they relate to natural philosophy. See above §3.4, p. 59 and n 144. The promise of a treatment of disease and health is repeated at the end of *PN* as we have it (*JSVM* 27 480b24-30), along with the thought that medicine and natural philosophy are close relations. If Aristotle did treat health and disease, the treatise has not survived. Cf. Hett 1936: 388. A work on the origins of *diseases* is mentioned *PA* II 7 653a8; cf. Flashar 1962: 318. We might compare the modern view that there is no point in talking about ageing in isolation from pathology, even though ageing is not a disease, since it is an increase in vulnerability to disease, i.e. a decrease in resistance. Cf. Comfort 1979: 31, 254: '[Ageing processes] contain the element of the unprogrammed, of progressive malfunctioning not found elsewhere outside pathology. In this case, however, the malfunction and program loss are "normal" in the sense that they are inherent in the system, and this is what ageing implies, ... the homeostasis ... that has brought the system through epigenetic development to adult or "normal" function becomes increasingly unstable; the variations to which it must respond become larger and its capacity to respond to them, less.'

85. Cf. 466b6: the capacity of residue is destructive of either a part or the whole. On seed as a residue cf. *GA* I 18 724a14-726a25.

86. Cf. Wardy 1990: 217 n 86 on *Cat.* 8 9a21-4, and 219.

87. Wardy 1990: 216-20; the evidence that Aristotle thought qualitative change necessary for becoming healthy: *dC* I 3 270a27-9; cf. also *Metaph.* VII 7 1032b18-28. Cf. the definitions of health *Ph.* VII 3 246b5, cf. also *Top.* 145b7 ff (in contrast with the beta version of *Ph.* VII 246b21 – see Wardy 1990: 216 f, and n 84; cf. also *Ph.* 210a20), Tracy 1969: IV a, 157-63.

88. 466b16-28, cf. *LBV* 1 465a7 ff.

89. *LBV* 3 465b27ff. On the environment cf. *HA* VIII 2, and above §4.2.

90. 466b21-2.

91. 466b28-33.

92. On a larger flame consuming a smaller one see p. 82 and n 39, 94 and n 119.

93. Balme 1992 ad *GA* I 18 724b23. On *syntêxis GA* I 18 726a21 ('always morbid'), 725a33 ('with no natural place of its own'); Theophrastus uses colliquescence to explain fatigue (*On fatigue* esp. 3, 5).

94. On the connection between wetness, food and heat see *Metaph.* I 3 983b23 and *Meteor.* II 2 355a4. On heat and concoction see §4.4 below.

95. The last part of the chapter (466b33-467a5) concerns further divisions that we met in chapter 4: aquatic animals are less long lived for being watery. This is presumably a development of the basic criteria for longevity discussed at the start of the chapter, since it leaves them cold and easily frozen. Non-sanguineous animals fall into the same class, since they have no fat or sweetness.

96. 466a9.

97. *LBV* 1 464b22-6; cf. Ross' title for chapter 6 (1955a): 'On the distinctive natural cooling of plants.'

98. One other difference between plants and animals deserves mention: plants do not have stomachs (*PA* II 3 650a21-3). The preliminary working up of their food takes place through the heat in the earth. Nonetheless, they have warmth in them, to do the rest of the working up of nutrition (*JSVM* 6 470a20). Their less obvious heat, in comparison to e.g. that of humans, might explain why there is more emphasis on their drying up. While animals grow cold and dry as they age, ageing in plants is in fact called drying up (*auansis JSVM* 23 478b28).

99. 467a7-9.

100. 466a20.

101. 466a29-32.

102. *liparotês*; cf. 467a4.

103. Elsewhere, cephalopods, *malakia*, are said to have a sticky nature (*GA* II 1 733a22 ff). Stickiness (*glischrotês*) is a derivative quality – i.e. of wet-dry, hot-cold

(*GC* II 2 329b32). There is a back reference at 467a18 to 467a11-12, comparing plants to animals.

104. 466b10-13. On the division of animals cf. *dA* I 4 409a9-10; 5 411b19-21; II 2 413b16-23; *JSVM* 2 468b1-15, above p. 69 ff.

105. *dA* I 5 410b23, II 12 424a33.

106. 467a12 following Siwek's translation (1963): 'plantae semper juvenescunt', and Ross (1955a): they are 'always renewing themselves'. For a strikingly similar phrase, cf. Plato's *Symposium* 207d7 (for the significance of Aristotle's use of this work cf. above p. 50 and n 98). For the unity of plants, see above §3.5 p. 70.

107. *LBV* 6 467a20-3. At 467a24-6 Aristotle notes that there is a 'small difference' between plants renewing their parts and being long lived through grafting.

108. *dA* II 4 416a4, *IA* 4 705a29-b8; *PA* II 2 648a12; *HA* IV 11 538a23. See Lloyd 1983: 94-105.

109. *PA* IV 10 686b2 ff (cf. Althoff 1992: 151, Freudenthal 1995: 57-9) would seem to suggest that dwarfs are cool, i.e. underdeveloped. Yet *LBV* 6 suggests they are long lived because hot. On dwarfs cf. also *Prob.* X 14, where their condition is attributed to a shortage of food; also *MR* 2 453a31-b7; *SV* 3 457a22.

110. *PA* II 7 on the brain esp. 652b7-653a10.

111. The processes described in *Meteor.* IV 1 occur both in whole plants and animals and in their parts (378b31 f). On this chapter cf. Freudenthal 1995: 13-14, 40-2, Tracy 1969: 174-8. The authenticity of *Meteor.* IV is generally accepted (Lee 1952: xiii ff, Düring 1944: 17 ff, 107 ff. Freudenthal 1995: 13 n 18), but for further doubts see Strohm 1983, who thinks it has been reworked after Aristotle.

112. Cf. Lee's note (1952: 291): 'rotting' (*sêpsis*) refers to 'cases in which a thing decays, disintegrates or perishes in the ordinary course of nature (cf. 379a3), its literal meaning being putrefaction.' Cf. also Alexander 1899: 181.27 ad 379a2.

113. 378b10-26, b26-379a11. 'Proportion' translates *logos* 378b33. Alexander (1899: 181.32-182.2 on 378a30 ff) comments that Aristotle means the four capacities in respect of which there is coming to be and passing away, in which capacities all plants and animals participate; cf. above §4.1.

114. IV 1 378b26-379a2, cf. 2 379b34 ff.

115. See *Meteor.* IV 2-4 for the specific discussion of the kinds of heat esp. 379b18-30 on concoction. On concoction see below §4.4 p. 104 f. Alexander (1899: 182.31-183.5, 185.13-23) emphasises the role of control and proportion (*analogia*). He also thinks that old age is *sêpsis*. Düring (1944) renders *gêras* (379a5) 'senile decay'; that is just what Aristotle is saying: old age is a kind of perishing. For form's loss of mastery over matter, cf. *GA* II 4 770b11-19.

116. *Meteor.* IV 1 378b28-379a6. It is instructive that *sêpsis* is not confined to things that are alive, since other things, such as sea water, are composed of mixtures which become unbalanced and decay (379b4-6). Not all mastering heat is the heat of living things.

117. 379a11. Cf. *GC* I 10 328b3: wet is *euoriston*. Loss of heat brings with it a loss of moisture, and that results in dissolution (*Meteor.* IV 1 379a23-6; cf. Freudenthal 1995: 125-6, 151-7). Unlike Düring (1944: 66), I find no discrepancy between the theories in *Meteor.* and *PN*. He quotes *JSVM* 23 478b32, where he takes the indefinite article to indicate uncertainty; but it indicates 'some kind of heat', i.e. connate heat.

118. *Meteor.* IV 12 389b28-390b2. Aristotle insists (390a3 ff) that functionality extends right down through the various matters in a living being. This functionality is, however, less clear (less determined) the further the analysis is pursued. What capacities of moist stuff achieve is not as clear, determined, as e.g. the function of the eyes.

119. *Meteor.* IV 1 379a16-18. Cf. Alexander 1899: 222.14-16. On internal vs. external heat cf. *Meteor.* IV 3 381b18, and Plato *Philebus* 29b-c and Freudenthal 1995: ch. 1. Cf. the destruction of a lesser fire by a greater one in *LBV* 3 (465b23 ff) and 5 (466b30-3), also *MA* 5 700a31-b3, *GC* I 7 323b8-10. Cf. also Alexander 1899: 185.5 ad *Meteor.* 379a19 on how the cold internal to the thing is the cause of decay: because heat draws in the wet from outside, that which is rotting becomes dry after becoming wet (15 ff). Hence there is more rotting in hot regions than in cold ones, since the environment is not hot enough in the latter to take control of the heat in each thing (18 ff). See also Freudenthal 1995: 155 on these passages; and Theophrastus *On fainting* (FHSG 345 – Photius' report), *On fire* 1, 10.

120. *LBV* 5 466b16-28, cf. 3 465b27.

121. See Comfort 1979: ch. 1.1 on difficulties to know what one is measuring when answering the question how long something lives. One central text for the question of Aristotle and measurement is *Ph.* VII 5, on which see Wardy 1990: ch. 8; cf. also Hussey 1991.

122. 467b5-9. Hett 1936: 388 ff in his introduction to *LBV* and *JSVM* says that both treatises are continuous treatments of 'how life is maintained in the organism and what the causes are which lead to deterioration and dissolution'. At 467b4 we are referred to a work on plants for a treatment of 'treelike' plants and annuals: on this work see Ross 1955a: ad *SS* 4 442b24-6, and Bonitz 1870: 104b38.

123. If one of the adjectives in the phrase *symphyton thermotêta physikên* at 469b7-8 is not a gloss, the question would arise what the difference is between being natural, and being connate heat. Related phrases are innate (*emphytos Meteor.* II 9 355b9), natural (*physikê* e.g. *PA* II 3 650a14), psychic (*psychikê GA* II 1 732a18) heat (*thermotês* or *thermon*); natural fire or the fire within (*physikon pyr JSVM* 14 474b12, *to entos pyr* 12 473a4). Ross 1955a: 41 notes the connection between this heat and the central organ and digestion; he thinks all these phrases refer to connate heat. Note that the adjective *zôtikos* means 'like, characteristic of or belonging to animals' (e.g. *Ph.* VIII 4 255a6), and it is in this sense that it is used of heat; thus translating *zôtikos thermotês* (*JSVM* 12 473a9 f) 'vital heat' may be misleading; one passage where such a translation may, but need not suggest itself is *GA* II 4 739b23. On vital heat, cf. Rüsche 1930: 192f.

124. Cf. 469a24, and 468a20-3 *kat' aisthêsin*.

125. The necessity involved here is hypothetical necessity: if the animal is to function well, then certain things must obtain; see below p. 114 f.

126. The well known texts on Aristotelian principles are *APo.* II 19, *Metaph.* I 1; cf. *APr.* I 30. For the natural scientist's need for experience cf. *EN* VI 8 1142a12 ff.

127. On final nourishment (*eschatê trophê*) cf. *PA* IV 4 678a6-20. See above §3.3: conversion of food is a gradual process.

128. Michael 1903: 108.23 ff. 'Plays' translates *kinei* which literally means 'sets in motion'. An important parallel to this passage is the Ps.-Platonic *Alcibiades Major* 129b-130c, where, as part of the argument that a human is really the soul, the soul is said to use, and so to be different from, the body. It is, of course, only a partial parallel, since for Aristotle the idea that user and used are different serves not to distinguish soul from body, but different parts of body, which play different roles within the actuality of nutrition.

129. Reading *kai ton topon* at 469b2, with most of the Mss., and Ross (1955a) and Siwek (1963).

130. As stated at the start of *JSVM* 1 467b16, above p. 65.

131. 469b13-17. Cf. Ogle 1897: 113 n 33. 'So called death', presumably because death is restricted to animals; plants dry out (*JSVM* 24 479b3); for Aristotle, both are a perishing of heat. Note that we do not have detailed descriptions of the structure of the heart in *JSVM* (Lloyd 1978: 232), merely a summary (JSVM 22 478b1 f) fitting it to its function; but of course, the accounts of the heart in *HA* and *PA* are connected with its function. For the conflicting accounts of the heart in *HA* III 3, *PA* III 4 and *SV* 3 see Lloyd 1978: 242-4; cf. Wiesner 1978, Platt 1921: esp. 522-3. On the difficulties with Aristotle's view of the anatomy of the heart see Ogle 1897, Thompson 1910: ad *HA* III 3; Platt 1921; Harris 1973: 121-76, esp. 160-73. Aristotle compares the heart to a hearth keeping the warmth as 'kindler of nature' (*tês physeôs to zôpyroun PA* III 7 670a25); cf. *PA* III 4 666b3 and 665b6-16, 666a3 on the heat preserving the heart; for the connection between the heart and life and death, cf. 667a33-b1: the heart sustains no strong *pathos* (affection) since when the *archê* (principle) perishes there is nothing to help it.

132. 469b16 *tês psychês hôsper empepyreumenês en tois moriois toutois* (i.e. in the heart and its bloodless analogue). Michael (1903: 109.8-9) suggests seeing this turn of phrase as a comparison: 'as fire comes to be alight and exists in the charcoal, so is the soul in the heart and its analogue'. The problem with this is that the soul is not merely like the fire in the charcoal, it also depends on heat consuming food. Contrast the phrasing in the summary in chapter 14 474b10-13: there is no nutritive soul without *physikon pyr, en toutôi gar he physis empepyrêken autên* (i.e. the nutritive soul); cf. also *JSVM* 22 478a30.

133. *PA* II 7 652b8-26, esp. 10-15: 'Heat is the most serviceable of the bodies for the work (*ergon*) of the soul. Nourishing and moving are the work of the soul, and these occur most through this capacity. Saying the soul is fire is like saying the saw or the drill is the carpenter or his art, because the work is performed by them both when close together.' Cf. *JSVM* 14 474a25-30, esp. 28: 'everything is worked on *by fire.*' This would seem to apply as much to food in an oven as to that which we digest. Concoction and nutritive soul are both mentioned in the summary (474a26-8, b10-11).

134. 469b8 f.

135. 469b13, 5 470a5-7. Michael (1903: 100.15-18) refers to *LBV* 5 466a20-2 (hot and wet are the matter of living beings) in his background account of why the heart is said to be in *JSVM* 4 the seat of control (i.e. of the nutritive soul) over the rest of the body: 'Such (i.e. hot and wet) is the place around the heart in which the nutritive soul is.'

136. Cf. §3.3 above. Contrast Freudenthal 1995: 126-30 who talks of the heart producing 'vital' heat, which then has to be distributed to the rest of the body using 'pneuma'.

137. *JSVM* 4 469b11-13. On concoction see below p. 104 f.

138. *GA* V 4 784a34-b3, but only one principle i.e. soul is needed (*MA* 10 703a34-b2). That is, there is one centre where nourishment for the rest of the body comes from.

139. *JSVM* 1 467b16; see above §3.5 p. 65.

140. Cf. §4.1 p. 77 f on Waterlow's view of the constituents of bodies.

141. Gill 1989b: 212-22.

142. Freudenthal 1995: 36-47.

143. See below §4.6.

144. Hett's (1936) translation of 469b18-20.

145. On this necessity, see below p. 114 f.

146. Fire: Ross (1955a: 303 ad 469b23) Biehl 1898; connate heat: Siwek 1963: 297 n 46, Althoff 1992: 154.

147. 474a26. Siwek (1963: 345 n 50) thinks the heat needed is not 'ordinary' heat, but pneuma. This is an example of a common confusion between connate pneuma and connate heat, and it will be necessary to show how they are related (see below §4.7). In the context of spontaneous generation, that is, generation of animals without progenitors, Aristotle sees himself forced to claim that *everything* is filled with 'psychic' heat, so that in a way everything is filled with soul (*GA* III 9 762a18-22). A weak, but feasible reading of this would be say that this heat is psychic in that it can, under suitable conditions, have the capacity for psychic function (nutrition).

148. *Meteor.* I 3 340b23. For the boiling of blood in the heart cf. below §4.7; flame is defined as a boiling of dry pneuma (*Meteor.* I 4 341b22, contrast *GC* II 331b25 and Joachim 1926 ad loc.).

Fire is said to be the only element which is always in a substrate (*GA* III 11 761b19 ff, cf. *PA* II 2 649a22), which Theophrastus discusses *On fire* 3. As Theophrastus says (*On fire* 4), this is connected with fire's need for nutrition (*trophê*); and of course a boiling is always a boiling of something. For a pun involving boiling and life, see *dA* I 2 405b28.

149. 469b21, 23 keeping the Mss reading (and Michael's 1903: 109.11), against Biehl 1898, Ross 1955a, Siwek 1963.

150. Connate heat is not fire and does not originate in fire (*GA* II 3 737a5-7). Balme's remark seems just (1992: 147-8): heat is a power primarily had by fire. On the similarity of fire and innate heat, I follow Balme 1992: 163-4, Althoff 1992: 173, 161, 65, as against Freudenthal 1995: 109-11. (Althoff 1992: 194, however, denies that this similarity holds in *GA* II 3 736b29 ff.) On the distinction between internal and external heat cf. *SS* 4 442a6-8 where they are simply divided by being in the living being or not. On heat in living things see *PA* II 2 648b11-649b13 on the ambiguities of the word 'hot'; 648b36 f on proper and alien heat: things that are hot in virtue of themselves and those that are not.

151. 469b21: *horômen* 'we see'; on the special status of the sun's heat cf. Theophrastus *On fire* 5.

152. My rendering of these terms follows the Oxford Translation (G.R.T. Ross 1908, also preserved in the Revised Oxford Translation, Barnes 1984).

153. See *JSVM* 14 474b18 f.

154. 469b29-31. In the Greek, the final words are: *prin epistênai tên anathymiasin.* LSJ s.v. *ephistêmi* B 3, suggest the verb means 'form', citing our passage, and apply it to the vapour: they think that this use of the verb also applies to the formation of cream on top of milk. However, it is not merely the vapour that is important, but the establishment of its continuous production; so it would be best to take the 'formation of vapour' as the continuous process of evaporation (cf. *anathymiasis* at *Meteor.* II 8 366a6). Other translations seem unsatisfactory: G.R.T. Ross (1908 Oxford Translation) offers 'before more [nutriment] is sent up by exhalation' i.e. he takes *anathymiasis* to refer to the process by which food is fed into the heart or fire (cf. Siwek 1963 'prius exaurit quam evaporatio ei opem ferre possit', and Althoff 1992: 154); however, it seems clear from parallel passages that it is the process of burning that is meant by this term (cf. *SV* 3 456b19 ff where the process of evaporation (*anathymiasis*) in the heart is contrasted with the result (*anathymiômenon*) and cf. *JSVM* 26 480a10). Hett (1936) translates 'before the evaporation comes to a stand', which gives little sense, as does Ross' (1955a: ad loc.) précis 'before it ceases to smoke'; Buchheim 1999: 29: 'früher als die Aufdünstung [neuer Nahrung] sich darauf aufbauen kann' is closest to my understanding of the passage. *Anathymiasis* is used of evaporation both in living things and in other contexts; contrast *Meteor.* I 3 340b27, 29 where it is used of dry evaporation (i.e. vapour) from the earth in the air, and is said to be potentially fire (cf. also *Meteor.* III 6 378a18 ff).

155. Michael 1903: 110. 1 ff, esp. 11-14 cf. above pp. 82 and n 39, 94 and n 119.

156. Siwek (1963 translation of 470a4-5) translates: 'ignis perpetua gignitur et fluit similiter ac flumen', but he thinks that the river here represents the natural demise of the fire: the flow balances coming to be. It seems, however, that a fire is merely coming into being, after all it does not perish into anything. As Mary-Louise Gill has pointed out to me, the water in a river is not perishing, it is merely flowing past.

157. This reminiscence is not noted by Marcovich 1967: 194-214 under his number 40 (= DK 22 B 12). It offers clear circumstantial evidence that Heraclitus Frg. DK B 12 is concerned with the soul. Kirk (1962: 367) doubts (against Diels-Kranz) that the reason the words are cited – to show that Heraclitus thought that the soul is *aisthetikê anathymiasis* (keeping Wellmann's emendation) in support of the Stoic view – is vouched for in words from Heraclitus himself (so too Marcovich 1967: 213 n 1). On the Stoics here cf. Solmsen 1968. That Aristotle so clearly alludes to Heraclitus when he is dealing with the soul (albeit not explicitly the perceptive soul) and evaporation (*anathymiasis*) is good reason to think that *psychai d' apo tôn hygrôn anathymontai* forms part of the quotation in B12, as Diels-Kranz think (cf. also *dA* I 2 405a24, quoted as part of DK 22 A 15, and used by Dilcher 1995: 63 as part of his defence of *anathymiasis* as a Heraclitean term); so too Buchheim 1999: 30 n 77. The connection with our passage disarms Kirk's main objection, viz. that there is no connection between the flowing of rivers and the soul. On change escaping our notice cf. *Ph.* VIII 3 253b9; on rivers, fire and air always becoming other *Meteor.* II 3 357b27. In a discussion of friendship, and how opposites can be useful to one another (*MM* II 11 1210a17), Aristotle (if he is the author) says that 'they say' a moderate amount of moisture feeds fire; Mansfeld 1967: 28 n 4 thinks this is a reference to Heraclitus. On Heraclitus' idea of burning and its use by Aristotle, cf. Buchheim 1999: 28-31 who independently reached a similar position to the one advanced here.

158. *LBV* 5 466b29-33, cf. *LBV* 3 465b23 ff, and Heraclitus DK B 36.

159. Contrast what he says about flame (*Meteor.* II 2 355a9-11): it comes to be when wet and dry change. It cannot, however, be *fed*, since it does not stay the same. One of the conditions growing things must fulfil is that they stay the same (*GC* I 5 321a22 – see above §3.3 p. 55 f), otherwise what happens is coming to be: a child's growth is not the generation of a new human.

160. 470a5-18. Cf. Michael 1903: 110.17-25. For Theophrastus' (different) explanation of the same phenomenon, see *On fire* 11, 23, 24, and Sharples in FHSG Commentary 3.1: 127 n 359.

161. See the discussion of Aristotle's arguments for the unity of the principles of perception and nutrition in *JSVM* 1, above §3.5

162. 425a6.

163. The status of these chapters has long been a question of debate. Lawson-

Tancred (1986: 217) sees the chapters as transitional to biology; they certainly lead on to the *PN* satisfactorily. Hicks 1907: 573 remarks on the purpose of *dA* III 12: 'It remains to consider the parts taken by the various faculties of soul in the maintenance of life; in other words why living things are found to possess one or more of these faculties and what is the end which each faculty subserves.' Such a treatment has been promised at III 2 413b9-10, 414a1, b33, but he thinks that the return to nutritive soul is somewhat abrupt (cf. Hamlyn 1993: 156): the chapter starts 434a22 with a 'therefore', but it is not clear what this follows on from. So perhaps some text is missing before the chapter. There may, however, be a connection with the treatment of appetite in the preceding chapters: if there is appetition, there must be nutrition. Any account of appetition presupposes the continual need for nutrition. Hutchinson (1987) thinks the chapters should come between II 4 and 5; however, he does not take account of the relevance of locomotion and appetite to III 12 and 13, which makes the earlier position inappropriate. The systematic interrelation of the faculties can only be discussed after the individual faculties.

164. 434a22-5; 434a32-b8. On nutrition and growth, see above §3.3.

165. II 2 413b4-7, 3 414b3.

166. This seems a reasonable reading of the 'organ within' in *dA* II 11 423b22 (so Hamlyn 1993: 111), in the case of animals with hearts; contrast *PA* II 8 653b24 ff on flesh as the medium or 'primary' organ of touch. For touch and heat, cf. *SS* 2 439a2-4.

167. 434b9-24.

168. *SS* 1 436b12-18.

169. 434a27 cf. II 11 423a13-21.

170. *PA* II 1 647a6-21. On flesh as a mixture, see *dA* II 11 423a12-15.

171. 435a21 ff. On the organ of touch being in a mean state in the ranges of tangible qualities, see II 11 424a2-15. Hot and cold are species of the same genus (*GC* I 7 324a1-5). The non-basic qualities, such as hard and soft, are composed out of mixtures of the four basic qualities (*GC* II 2 329b7-26). It is very contentious quite how the matter of the sense organs is involved in perception; see especially Sorabji 1974, 1992, Burnyeat 1992, Webb 1982, Sisko 1996, Van der Eijk 1997, Johansen 1998. On the mean in perception cf. Burnyeat 1992: 20-1, cf. Tracy 1969: 212-20. See below §4.7 p. 123 for the chilling involved in palpitation, and cf. e.g. *PA* II 4 650b28-32 on cooling in fear.

172. *Meteor.* IV 4 382a17-21.

173. *dA* II 11 423b27-424a2.

174. 435a21-5.

175. 435b2-5.

176. 435b4-25, cf. II 12 424a28-32, III 2 426a27-b7. For a defence of the relevance of these arguments to Aristotle's understanding of perception, see Sisko 1996. Barker (1981) sees a very restricted role for the mean or ratios in perception.

177. *GC* I 5 322a10-16.

178. *GC* II 2 329b26-29.

179. *Meteor.* II 2 355b6-11. On the polyvalence of concoction in Aristotle, see Lloyd 1996: ch. 4. Everything is nourished by the sweet, either alone or in combination. (*SS* 1 436b15-18, 4 441b27-442a12. This text then refers to *GA* III 11 762a12-13, V 1 776a28-29 – Ross 1955a: 206 ad *SS* 4 442a3.) On residues cf. above §4.3 p. 89 and n 83.

180. On this term in *JSVM* 4 469b11-13, see above p. 97 and see Lloyd 1996: ch. 4.

181. Reading *pepainomenon*, a gloss in O, cited by Thurot according to Lee 1952.

182. *Meteor.* IV 2 379b18-30 (trans. Lee 1952).

183. See above §2.2 on natural form as the result of coming to be, and on food as matter see p. 25.

184. This interpretation of extinction and exhaustion of heat is confirmed by the expressions used in the resumé in *JSVM* 14 474b14 ff, esp. 19 ff: 'Extinction happens through an excess of congealed cold. ... (*pêgnoumenou dia psychous hyperbolên*) and animals die .... Exhaustion occurs through the amount of heat. For both if the surrounding heat is excessive or if it takes no nourishment, the burning thing perishes' (*hê de maransis dia plethos thermotêtos. kai gar an hyperballei to perix thermon, kai trophên ean me lambanei, phtheiretai to pyroumenon*). On balance, see further §§4.5, 4.6. On the question whether exhaustion is natural demise, and extinction violent demise (469b23), see below Chapter 5.

185. The brain is notably absent from the *PN* account of the preservation of a balance between hot and cold in the whole body (contrast *PA* II 7 652b7-27), although the balancing role of the cold brain is mentioned (*SS* 2 439a3, *SV* 3 457b26-458a4). The reason might be that the brain, unlike the heart and lungs, does not engage in any obvious action; and the *PN* is about the *praxeis* of living things (see above §3.1).

186. 470b1-5. Cf. the resumé of the early part of the work at *JSVM* 14 474a25-b9.

187. 470a25-7 cf. *PA* III 4 667a27-8, *JSVM* 15 475a11-15 and the reference forward to *JSVM* at *SV* 2 456a6-11.

188. Cf. *SV* 3 457b18-19.

189. Plants feed on a mixture of earth and water (*GC* II 8 335a13 ff, *GA* III 11 762b9-12, *PA* II 3 650a1-6). See the definitions of hunger and thirst in *dA* II 3 414b11: hunger is for the dry and hot, thirst for the wet and cold; for fire feeding on the wet see *Meteor.* II 2 355a3-5, on taste *dA* II 10, *SS* 4.

190. 470a27-32.

191. Michael 1903: 112.1-4.

192. Note *hyperballei* 470a27. On balance see above p. 105 and n 184.

193. *PA* II 3 650a21-3.

194. See *GC* II 10 336a31-b15 and Williams 1982: 191.

195. 470a32 ff. For the idea that the organs of plants are 'thoroughly simple' see *dA* II 1 412b1 f; plants have few actions (*praxeis*) to perform, and so need few organs (*PA* II 10 656a2).

196. *JSVM* 7 esp. 470b6-9. This emphasis on function is also to be found in *dA* III 9 432b9-12, which refers to the necessity of dealing with inhalation and exhalation, in giving an account of growth and decay.

197. 470b10-12. If one wishes, one may imagine that contemporaries had made the complaint, common in modern discussions of Aristotle's methods, that he treats his predecessors high-handedly, when he construes their positions as a way of arriving at his own.

198. For a brief discussion see Althoff 1999: 78 f.

199. 471b23-26, cf. 10 472a2.

200. See e.g. *JSVM* 25 479b15, 27 480a20. This was also the point of quoting the lines criticising his predecessors in our discussion of the status of ends for Aristotle (see above §2.3 p. 30).

201. *JSVM* 9 471b19, cf. 7 470b9-10.

202. *JSVM* 1 467b12, cf. 10 472a26.

203. *JSVM* 15 474b30.

204. Cf. the use of classification in *LBV* 4, see above §4.3 p. 86.

205. Cf. 474b26 f.

206. 474b30-475a15, 20-5. Animals that are both sanguineous and have a lung, but a spongy, rather bloodless one, can survive with rather little breathing, since their lung is not so bloody, i.e. hot (475a20-9). They can even survive for some time without breathing, but finally their heat will consume itself, as has been explained in *JSVM* 5: this kind of violent perishing is suffocation. 475a 20-9 does connect with what went before a15-20 (*pace* Ogle 1897: 123 n 89 ad loc.), and is not primarily about bloodless animals: the point is the similarity between animals without blood, and those with only a little: they can live for a while without breathing. In the next section 475a29-b14, the topic of suffocation, introduced here, is taken up in the case of insects and water animals. Ross (1955a: 321) also thinks a20-9 in place here but sees the link in longevity of some blooded animals and insects; but the length of life in question is that possible without breathing. Siwek (1963: 346 n 72) finally also sees the lines as in place, but his reasons do not emerge clearly.

207. On the supposedly embarrassing situation caused by the longevity of trees, see Freudenthal 1995: 190-1. There is, however, no sign of Aristotle being forced by any systematic commitments here.

208. Following Bonitz 1870: 668b30, 46 ff s.v. For the first meaning see e.g., *Metaph.* X 1 1052b29, for the second *PA* II 7 652b17. Bonitz distinguishes between downward tendency, 'insita vis aliqui rei movendi', and a third meaning (b55), whereby *rhopê* is used as a metaphor for *dynamis*. (He lists (b40) the use at 474b31 as being transferred from the first meaning, along with 478a16, which cannot be right.) At 652b17, in a passage discussing why hot animals need the cold brain (in Aristotle's view), everything is said to 'need the opposite *rhopê*, so that it achieves moderation and the mean'. On 474b31 cf. Siwek (1963: 321 and n 62); the same image is used below 478a16 for very hot animals: the fire is so hot that it is always on the point of going out by exhaustion, so requiring elaborate cooling to remain in balance (Ross 1955a: 331 ad 478a15-18, Michael 1903: 140.22-3, cf. p. 119). The two texts seem confused – both cool animals and hot ones have a small momentum (*rhopê*); but remember Aristotle's ideas about burning: it can go out through an excess both of cold and of heat. The use of *rhopê* is widespread in Aristotle's natural philosophy. The image of scales for the relation between life and death is old and common (see LSJ s.v. *rhopê*: most famously, Homer *Iliad* 8.72, 22.212, but cf. also Sophocles *Oedipus Rex* 961). For related images cf. *EN* I 10 1100b25, *Pol.* VI 6 1320b33, cf. Plato, *Republic* VII 556e2. The scales on the 'Boston Throne' may also belong here, but see Mertens-Horn 1997: esp. 102 (I owe this reference to John Herrman). On weight and balance, cf. Hussey 1983: 189, 1991: 223-6 ff.

209. Michael 1903: 128.25-7 ad 474b25-475a3.

210. Michael 1903: 131.23-132.19 ad 475a29-b5.

211. Against Furley's view (1984: 14) that breath in Aristotle's view is 'some kind of substance whose consumption is necessary for vital heat'. We return to this point *in extenso* below, §§4.6 and 4.7.

212. Ogle 1897: 123 n 94. quoted by Ross 1955a: ad 475b5-14.

213. *JSVM* 10 472a31-5.

214. *JSVM* 11 472b29-35; at 15 475a7 he compares the buzzing of insects to panting, as being made with breath.

215. The function of the lung: *PA* III 6 668b34-669a1. As Lloyd remarks (1992: 149), for Aristotle all biology is related to soul.

216. He refers to *JSVM* ('in our discussion of respiration' *en tois peri anapnoês*, always on the subject of gills) at *PA* III 6 669a4-5, IV 13 696b1, 697a22.

217. Cf. 470b13-15.

218. Newts, with legs and gills, cause some problems for the division land animal/water animal (476a5 f cf. *HA* VIII 2 589b26 ff).

219. On double function, where two organs are impossible, see *PA* IV 6 683a22-6, II 16 659a21-3, III 1 661a34-b5, b17-19, *dA* II 8 420b16-22.

220. 477a13 f refers back to 470b12-13; For the terms 'inhalation' (*eispnoê*), 'exhalation' (*ekpnoê*) and 'breathing' (*anapnoê*), see *HA* I 1 487a29, 11 492b6, *PA* III 3 664a19; 'breathing' can mean either breathing in, or the complex activity 'breathing in and out', which, Aristotle insists, must belong together (*JSVM* 8 471a7 ff, cf. also *SS* 1 436a15). According to Aristotle, Anaxagoras and Diogenes (470b31) and Democritus (470b28), all thought that air entered the body even where no lungs are present. Empedocles is said not to have answered the question (473a16).

221. It is unclear what animals with lungs are being compared with at 477a18: are they more developed than plants, as the Mss followed by Siwek (1963), have it, or than cold things (Ross 1955a), or than fish (Biehl 1898)? (Michael 1903: 136.27-9 had 'animals' in his text, which is impossible, as he notes.) The emendations are not very convincing. Nonetheless, we seem to need a contrast between animals with lungs and those without. What the reading of the Mss gives us is a contrast between animals and plants; this would fit with the idea that greater heat goes along with a more honourable soul, that is, in the case of animals, the perceptive soul. There is nothing in Aristotle to suggest that animals with lungs have a different kind of soul to those without. Probably the text should be left, awkward as it is (with Siwek 1963: 350 n 114). The idea behind the Mss reading may be that hot animals with lungs have their upper part in the same direction as the cosmos, unlike plants which are upside down: note the concentration on the idea of natural places in the whole chapter. Part of the contrast with plants is that they have their heads in the unnatural position of downwards, instead of the more cosmically correct humans. Cf. *JSVM* 1 468a5 ff: man has the part by which he ingests food uppermost in the most conspicuous fashion among all living things, since his upper part lies in the same direction as that of the whole. Contrast the close connection between upright posture and thought at *PA* IV 10 686a27-31. Light upper parts and heavy lower parts are necessary for all bipeds' locomotion, because only that way do they stay upright (*IA* 11 710b5). On the hierarchy among animals in Aristotle, see Lloyd 1983: 35-43.

222. 477a15-17.

223. For the relation between heat and soul in nutrition see above §§3.3, 4.4, and on heat and perception p. 102 ff.

224. For the connection between intelligence and purity see *PA* II 2 648a10; purity is presumably not unconnected with heat in that the concoction by heat separates the different parts of the food. At IV 10 686a24 ff upright posture is put down to man's divine substance (*ousia*). In turn, this is connected to his intelligence, which requires mobility in the relevant part, and so an upright posture. Thus this explanation will also, one might think, involve heat.

225. 477a23-5 following the Oxford Translation (G.R.T. Ross 1908): 'Hence this organ as much as any other must be assigned to the essence (*ousia*) of the animals both in man and in other animals.' Similarly Ross (1955a: ad loc.) in his précis: 'contributes to essential nature' rather than Hett (1936): 'The reason for its (the lung's) existence (*ousia*) both in this and in other animals must be assumed, just as in the case of other parts'; so too Michael 1903: 137.4-12 ad 477a19-20, 23; this reading is supported by *PA* III 6 669b8-12. Substance here must be formal, in that the form is what the lung is for.

226. Ross 1955a: 328 ad 477a25-7.

227. 'Such things as do not come about for the sake of something but from necessity and through the moving cause', *GA* V 8 789b18, and cf. the beginning of the discussion V 1 778a34-b1.

228. See *GA* V 1 778a32.

229. ad 477a25, Michael 1903: 137.17-18.

230. Aristotle's phrasing here (*tên ex anagkês kai tês kinêseôs aitian*) strongly suggests he is taking the necessary cause and that of change together. Siwek 1963: 351 n 120; ' "Causa necessaria" ... in rebus naturalibus est earum "materia" et huiusce materiae motus (*Ph.* II 9 200a30)'. The present passage does not preclude the effects of surroundings not as it were ingested as matter: seasons and such like may well count as factors here (see above §4.2).

231. *SV* 2 455b26 f.

232. See above all *Ph.* II 9, *PA* I 1 639b21-640a9, 642a1-13, 31-b4; cf. also *Metaph.* V 5 on necessity; Sorabji 1980: 155 ff; Charles 1988, 1991: as he puts it: given p, necessarily because of this q (on his reading it characterises both efficient and teleological explanation, in contrast to Cooper 1987). The point is that the occurrence of the antecedent condition is not enough to explain the occurrence of the outcome; it is not the right kind of cause. In our case, a suitable environment does not bring about living things with lungs. Cooper does not mention our passage, but he discusses passages such as *GA* II 6 742a16-b17 (256 with n 13) on the heart and its relation to the definition of the living thing in contrast to hypothetical necessity, with reference to *PA* I 1 640a33-b1 where the possession of certain parts is related to what it is to be a man. But I think this passage in *PA* I must refer to hypothetical necessity, insofar as the only other possibility is the necessary interaction of materials (see 639b34). Cooper cites *Ph.* II 9 200b4-8 for his view that the necessary thing is in the definition; 200b6 (cutting will not take place *if* the saw does not have such and such teeth) makes clear this is hypothetical necessity. The point in *GA* II 6 with the heart, and in *JSVM* 19 with the lungs is that there is hypothetical necessity, of a particularly intimate kind; if there is to be a man, then there must be a heart. For the need for the central organ for life, i.e. existence of the thing concerned, see above §§3.5, 4.4.

233. See above pp. 81 f. This necessity, applying to the interaction in matter, is rightly defended as Aristotelian by Cooper (1985, 1987: esp. 256, 264).

234. Michael 1903: 135.20-4, 136.19; cf. also 137.10-11 ad 477a26: they have lungs because they have the most heat.

235. For the complementary relation between teleology and moving or material causes, see Sorabji 1980: 152 f, 162 f. Cf. also Barnes 1994 on *APo.* II 11 94b27.

236. *GA* II 1 731b18, cf. generally *APo.* II 11 94b27-95a9, and *PA* I 1 642a2-13, where the 'for the sake of which' (*hou heneka*) and hypothetical necessity are distinguished. At *PA* I 1 642a31-b2 he illustrates, none too perspicuously, his theory of necessity in living things using breathing (although not using his own theory of breathing, as noted by Cooper 1987: 257), see also Sorabji 1980: 155 ff, Kullmann 1985: 218 f.

237. Cf. §4.2 above.

238. 477a25-31.

239. At 477b12, Aristotle refers to *PA* II 2 648b2 ff for the discussion of which animals are hot and which cold. We are dealing with balance in the constitution of living things: for excess and deficiency, see 477b2 ff, 15 and cf. b29.

240. *LBV* 5 466a21 ff.

241. 477b16: 'Nature is preserved most of all in its proper places.'

242. Michael 1903: 138.26, 139.26-8.

243. G.R.T. Ross' translation of 478a1-7 (1908).

244. Ad 477b16 f, Michael (1903: 138.32-3) remarks, the proper place preserves the elements (cf. *Ph.* VIII 3 253b33). Yet this remark does not apply only to elements but also to living things. On the different relations between things and their surroundings in respect of food cf. *HA* VIII 2. It is difficult to see exactly which correlations between elements and kinds of animals are meant at 477a27 f. See Ross 1955a: ad loc; for animals at home in fire, Aristotle went to the moon (*GA* III 11 761b15-23; cf. *HA* V 19 552b16-18). The topic is taken up by Theophrastus (*On the causes of plants* I 21.5).

245. For the idea of having little momentum in *JSVM* 15, see above §4.5, p. 110 and n 208.

246. 478a30. On the 'kindling' of the soul cf. on 469b13-17 above §4.4 p. 97 and nn 131-2.

247. *syntrêsis* is used here for the connection; at *HA* I 16 495a25 for that between nostrils and mouth. The other references in *HA* are II 17 507a5-10, III 3 513a35-b1, 18-25. See below p. 127 f and n 287.

248. Ogle 1897: 131 n 132 ad 478b26 ff.

249. Cf. Ps-Aristotle *de Spiritu* 2. Debru 1996: 140 n 32 suggests that exhaled breath arises from consumed food in Aristotle's view; but there is no sign of this in the text.

250. For Theophrastus' different explanation of suffocation, see *On fire* 24.

251. Ross 1955a: 338 précis of 480a16. Breathing is an involuntary motion (*MA* 11 703b3-14, Debru 1996: 44 ff; cf. also *EN* III 1)

252. Chapter 26, 480a2 f, discussed below §4.7 p. 127 f.

253. There seem to be no very informative pictures of bellows surviving (but see Healy 1978: 194 and figs 50 and 51). The double bellows here would seem to be a simple form of multiple bellows, which were used in order to ensure a continuous blast of air. Such a continuous stream of air is not analogous to the cycle of inhalation and exhalation. Bob Sharples has suggested to me that double bellows would be like Ctesibius' pump; for a picture see Lloyd 1973: 102, fig 17.

254. For the detailed physiology of lungs see *PA* III 3 664a26 ff, 665a16, III 6 669a13-b12.

255. G.R.T. Ross' translation of 480b9-12 (1908).

256. Michael 1903: 143.30-2 ad 479b9, cf. *JSVM* 21 478a18-21. On suffocation see also *JSVM* 15 475a22-9.

257. Michael 1903: 143.33-144.3. Cf. Ross 1955a337 ad 479b17-480a15.

258. Tracy 1983: 327 on *dA* I 1 403a31-b1.

259. 479b19-26. Palpitations caused by hope and fear are unique to humans because they alone have a sense of the future (*PA* III 6 669a18-22). Cf. the account of sensations, the heart and hot and cold given in *MA* 7-8 701b16-702a21.

260. Cf. *GA* V 2 781a25.

261. Also rendered 'pneumatisation', 'aerification' or 'aeration'. Ross (1955a:

336 in his précis of 480a2) describes it as the 'aerification of the fluid as it is being heated'. Cf. also Peck (1953: Intro. §63, and Appendix B on connate breath, §31) For *pneumatoun* cf. *dC* III 7 305b14: the wet is pneumatised by the hot (Althoff 1992: 169-70); important for *JSVM* is perhaps the fact that this passage makes clear that Aristotle knew heating liquids makes the resulting vapour take up more room. This would then explain the passage of pneumatised blood into the rest of the body. *GA* III 4 755a19: yeast grows because 'its more solid portion turns fluid, and the fluid turns into pneuma' (Peck 1953, but in a footnote he also suggests 'becoming inflated with pneuma', referring to 762a19). The text continues: 'This is the handiwork of the natural substance of the soul-heat in the case of animals, of the heat of the humour blent with it in the case of yeast' (Peck's literal rendering from his footnote); II 3 737a11: the body of the male seed 'dissolves and evaporates' (Peck's translation). 'Evaporation' here should be taken to mean 'turns into vapour or pneuma'; that is to say, in pneumatosis, pneuma is not introduced from outside into a liquid, but results from heating. Blood is made hot through concoction: *PA* II 1 647b2-7, 3 649b20-7, 3 650a2-5; but it is not clear to me if blood only transports heat because it transports pneuma (Hübner 1999: 15 n 40 referring to 479b26-480a15). Rüsche (1930: 224) identifies pneumatosis with the 'evaporation' (*anathymiasis*) of blood.

262. 479b26-480a2 – note the interest in the contrast and similarities between sickness and health cf. above §3.4 p. 59 n 144, §4.3 p. 89.

263. See above p. 99.

264. That is, not clearly and closely related to capacities for change. Solmsen 1957: 120: 'vital and animating force'; Jaeger 1913: 44: 'the driving force of the nutritive soul'. Jaeger presents the most extreme of what might be called the vitalist view of innate pneuma. He calls it 'beseelter Lebenshauch, der die Organismen durchdringt und sie spezifisch vom toten starren Dasein des Steins und des Metalls unterscheidet' (1913: 43). On Solmsen's view of innate breath see Freudenthal 1995: 107-9, 109-11, who also discusses Balme's view. Balme (1992: 160-4), Freudenthal (1995: 111) and Ross (1955a: 40-3) take innate breath to be a material factor.

265. Ross 1955a: 40-3, whose account of the distinction between innate breath and innate heat I follow. Contrast Peck on innate breath (*symphyton pneuma*) 1953: 578-93, Appendix B: esp. §§ 31-2. While he thinks that pulsation is the clue to the upkeep of connate breath, and nowhere suggests it is connected to breathing, he does think that breath is already in the fluid, which acquires a 'special quality' by being in contact with heat. Also Nussbaum 1978: Essay 3; Jaeger 1913: 50 n 1, Freudenthal 1995: III.2. *JSVM* 26 and 27 do not talk of connate pneuma; nonetheless Torraca (1958: 228-9) and Nussbaum (1978: 375 f) think it possible that discussions of connate pneuma in *MA* 10 and *GA* V 4 could refer to the description of pneumatosis at 479b26-480a15. *MA* 10 703a9-10 talks of the preservation of connate pneuma, putting it on a par with other parts of the body (703a16 f).

266. *pros tên thermotêta tên entos dA* II 8 420b20; cf. also *SV* 2 456a8-10: the lungs and gills serve the preservation of the heat in the heart.

267. On the tempering of heat, see above §§4.4, 4.5. Jaeger (1913: 50 n 1) sees an identification of innate heat and innate breath at *GA* II 3 736b29, which is, however only an exaggeration, which is then corrected by Aristotle to mean: warmth is the *hê en tôi pneumati physis*, which Jaeger takes to mean that pneuma is the subject of heat. This is (partly) right, and can be taken to mean simply that breath is hot.

268. But breath is hot air (*GA* II 2 736a1), and how can hot air cause cooling?

The solution to this problem lies in the fact that such qualities as hot are comparative as between the elements: air is hot compared with water and earth, and not in comparison with fire (*GC* II 3 330b4, Ross 1955a: 42).

269. *JSVM* 15 474b31-475a9. They also use innate breath for smelling (*PA* II 16 659b14-19).

270. Force for animals: *SV* 2 456a16, *Pol.* VII 17 1336a38, *MA* 10. Formation of organs: *GA* II 3 736b33-737a1.

271. Sweet and potable matter in food is drawn by innate heat into flesh and other parts of the body, bitter and salty is excluded (*Meteor.* II 6 355b6-11); see above §4.3 on residues, p. 89.

272. 456b2-7.

273. *SV* 3 458a25-8; cf. above §3.5. Instructively, in its approach to nutrition this passage contrasts starkly with *dA* II 4 (above §3.3), but is similar in its interest in physiological process with *JSVM* 27.

274. 456b19-28.

275. 457b17-19 and on waking 458a24-5.

276. See above p. 101.

277. The first scholar in modern times to have made suggestions along these lines was Jaeger apropos *MA* 10 esp. 703a23-8 (Jaeger 1913: esp. 45, 48). Both Nussbaum (1978: 146, 158 f, 161) and Freudenthal (1995: 137-8) see this text in the same light as Jaeger.

278. Cf. above §4.1 p. 77 f; Freudenthal (1995: 142) admits that this is not a very attractive theory; he also sees in pneuma the 'privileged agent assuming the function of the nutritive soul' (1995: 139). One could agree with this if it meant that breath cools the heat down so that it lasts longer. Freudenthal only mentions air that is breathed in, i.e. breath, once, in order to distinguish it from innate pneuma (1995: 147). As is clear from *JSVM*, breathing is crucial to survival. The dispute about connate breath is thus whether the function it performs in the living thing is comprehensible, or apparently magical. Pneuma is said to be a tool for nature at *GA* II 3 736b30 ff. Freudenthal does not use this text to support his position with regard to *MA* 10, which is otherwise like Nussbaum's. His examination of the *GA* passage is very attractive: the analogy with the stuff of the stars is functional: breath performs the function of allowing sublunary things to perpetuate themselves, because of its role in the formation of the embryo. On this reading, which I find convincing, breath is hot air. Nussbaum (1978: 161 f) finds it difficult to accept this idea (*GA* II 2 736a1), and thinks it has to act more like a fifth element. (Althoff thinks heat sometimes behaves like an independent substance, not merely a quality, 1992: 185, 189). For air, if left to its own devices, would move to just below fire, leaving the unity of the living being destroyed; other problems arise from the explanation of motion: hot air could not expand and contract without constraint (cf. *MA* 10 703a19-25) or escape qualitative change.

279. 480a15.

280. Michael 1903: 145.13, 27-31. Freudenthal (1995: 121-2) has a similar idea about pneumatosis to mine, but he thinks the best parallel is boiling milk which, because of its chemical structure, retains the bubbles; I think that the use of the term *anathymiasis* shows that we should be thinking rather in terms of a vapour. Cf. Freudenthal 1995: 120 on 479b31: 'From a physical point of view the phenomenon [sc. of pneumatosis] is essentially the same as the formation of vapour through boiling inasmuch as boiling is due to the volatilisation ("pneumatisation") of fluid by heat.' We must take his qualification ('from a physical point of view') with a pinch of salt: he thinks that there must be some other psychological account alongside the 'physical' one.

281. Cf. Peck 1952: Appendix B§§31, 32, Introduction §63. 'Motion': as we have seen the local movement is connected to the expansion of the blood through the effects of heating.

282. On *anathymiasis* as the process of evaporation in *JSVM* 5 see above §4.4. Aristotle's evidence for food being concocted in the heart (480a7-9) is that in generation the heart contains blood before the veins are differentiated (cf. *GA* II 4 740a17-23, referred to here).

283. Cf. 480a2-13. Blood is sometimes thought to transmit sensations from the peripheral sense organs to the central sense organ. For a discussion, see Van der Eijk 1994: 81 ff. One problem with the idea of blood being transported to the centre is that Aristotle only implies it in one or two passages, and nowhere suggests that it is a constant phenomenon (on blood in general see *HA* III 19). The main passages are *SV* 3 456b18-28, *Ins.* 3 461a5, 461b11. Van der Eijk (p. 85) thinks that the latter passage, although not clearly in favour of blood as the transmitter, at least does not speak against it. Where blood moves to the centre, this movement is part of the recession of heat from the extremities incident on sleep: the fumes from concocted food (presumably blood) rise to the head, are there cooled, and so fall back to the heart where they dull its motions. Another problem would be in the speed of the return of the blood to the heart which would have to be very fast to account for the quickness of perceptions (461b13, 18-19). And a final difficulty would lie in motivating the normal flow of blood back to the heart: its normal course is to flow out from the heart to feed the body. Van der Eijk (p. 83 f) suggests that if blood acts as a medium for transmission as e.g. air does for sight, then local movement of blood would be unnecessary. The problem is that there is no obvious quality of blood suiting it to such a role, as transparency is in air. Webb 1982 suggests pneumatised blood as the transmitter of sensation. The main weakness in Webb's position is that he thinks that where heat is mentioned it must be in pneuma, which he sees as an organ, and the substrate of the psychic fire. Heat must be in all the mixtures of the body, and while the pneumatisation of blood into food involves heat, this does not imply that the heat that does this, the psychic fire, is itself pneuma or in pneuma. The passages Webb cites are *SV* 2 456a1-24, *PA* II 16 659b18-19 (in bloodless animals pneuma is responsible for smell), IV 13 697a27, as well as the last chapters of *MA* (no blood is mentioned there). No connection is made by Aristotle in any of the passages Webb cites between blood and pneuma; the only passage for this is *JSVM* 26 on the pneumatosis of the blood in the heart, where sensation is also mentioned but not in connection with pneumatosis but palpitations. Nonetheless, these changes are chillings of the heat in the heart, and so at least incidentally connected with pneumatisation. So perhaps the presence of connate pneuma in the principle (*MA* 10 703a14-28) is connected with the pneumatosis of blood in the heart. But quite how pneumatised blood would serve as a transmitter remains obscure.

284. On the two grades of nutrition, cf. *SV* 3 458a10-15; *SS* 4 441b28. Althoff (1997: 356-7) comes to the conclusion that food involved in nutrition and that involved in growth cannot be simply divided between wet and dry for nutrition, hot and cold are responsible for growth: heat is, of course necessary for nutrition. On separating through heat, see *GC* II 2 329b27.

285. Furley 1984: 19. On pulsation, cf. Althoff 1997: 359 with n 28.

286. See above §4.6 p. 119 f. Contrast Michael 1903: 146.34-147.10, 20-7.

287. For a similar reading of pneumatosis here cf. Althoff 1992: 169-70. Boylan 1982: n 42, says of *JSVM*: 'Aristotle does not mention the mixing of this air [from the lungs] and blood, which might have served as a source of pneuma', and suggests that in *HA* a path from the trachea to the heart is hinted at (I 17 495b10, 496a30). The second passage in *HA* denies that there is a joining passage between air passages in lungs and blood vessels in the lungs; and says instead that the blood vessels receive breath (pneuma) by 'contact' with the air vessels, and so pass it to the heart. That this detail is omitted in *JSVM*, and that other details here strongly suggest that no breath is consumed in breathing make one think at the least that the import of the introduction of breath into the body was not uppermost in Aristotle's mind when writing our text (despite the reference to *HA* at 478a26). Rüsche (1930: 213 f) thinks that the accounts here and in *HA* fit together, but admits that breath does not have to enter blood, since it leaves the body again. One may well surmise that although in the dissections he was right about how breath gets into the blood, and so heart, it was not necessary to his view of vital activity, and so is ignored here.

288. *JSVM* 26 480a2-7, 27 480a23-b6.

289. Cf. on fire and connate heat above §4.4 p. 98 f.

290. On necessity see above pp. 81 f, 114 f.

291. Cf. Ogle's 1897: 41 remark on breathing: 'The vital heat is made to be its own regulator; for it is by it that the mechanism is set in motion by which all excess is prevented and the heat kept within limits.'

5. Defining the life-cycle

1. Cf. esp. *APo.* II 2, 11, and on causal definitions see e.g. Le Blond 1979, Sorabji 1980: esp. 195-205, Guariglia 1985.

2. Cf. *APo.* II 10 on the kinds of definition and Barnes 1994 ad loc.

3. Why these definitions are not the very end of the enquiry is not entirely clear; specific points about breathing and related phenomena remain to be settled in chapters 25-8, which do not affect the general definitions. See above §4.6. *PN* offers other notable definitions as results: memory at *MR* 451a15-16, sleep at *SV* 3 458a28-32. See above §4.3 p. 94 for a suggestion as to why longevity is not defined in *LBV*.

4. Death is spoken of as *teleutê kai phthora* 'the end and perishing' at 479a32. Aristotle restricts death (*thanatos*) to animals, which is parallel to drying out in plants (479b3 cf. 478b28 where old age is parallel to drying out); cf. *JSVM* 4 469b19: 'so-called death is the perishing of the heat in the heart.' See above p. 97 and n 131.

5. 478b25-8. Contrast the popular idea that there are many ways of dying (perishing), but one way of being born (coming to be).

6. Aristotle discusses the question of which changes happened naturally as well as unnaturally in *Ph.* V 6 230a18-b10.

7. 478b31-3. This is quite generally true – not merely of living things (cf. *Meteor.* IV 1 379a16 above p. 93 f).

8. 478b30-3. On mature beings cf. *JSVM* 2 468a13: there are three parts to the complete living thing. The point is presumably that some incomplete things do not have their own way of taking or digesting nourishment.

9. 469b18-20.

10. See above §3.5.

11. Contrast the treatment of Democritus in chapter 10 472a16-20.

12. Cf. *Ph.* IV 8 215a1-3. Cf. *LBV* 3, where the question of whether perishing could be avoided by material things is discussed, perishing was divided into that which was caused from outside the living being, and that which was caused by residues within it (see above §4.2).

13. In *JSVM* 22, we have already been promised an account of dying *dia pathos ê dia gêras*, 'owing to disease or old age' (Hett's 1936 translation of *JSVM* 22 478b20). Ogle's (1897: 13) definition of violent death is that it happens 'from such adventitious cause as is not inherent in the normal constitution of the body'. This would in fact allow disease to be subsumed under violent death. Although it is internal to the living thing it is against its nature, and so violent. Michael also sees a reference to disease here and suggests inflammation of the lungs (*peripneumonia*) as an example (1903: 142.5 ff). Cf. the promise of a treatment of health and sickness at *LBV* 1 464b32-3.

14. 479a8-10; cf. also *LBV* 5 466b29-32, where the same verb *syntêkein* 'consume (itself)' is used as in 479a10, rather than *marainesthai* 'exhaust' as in chapters 5 and 24 (479b2). Cf. *Prob.* I 5, VIII 9. I am grateful to Bob Sharples for these references on consumption.

15. *geêra* not *gêina* i.e. 'earthy', not 'earthen' (cf. Bonitz 1870: 147a16; 154a49 s.vv.); 'earthy' means behaving as earth does (*Ph.* VIII 4 254b22). Cf. also the pun on *geêros* and *gêras* 'old age' *GA* V 3 783b7. The earth here is not playing the role of proximate matter as envisaged in *Metaph.* VII 7 1033a5-22, IX 7 1049a18-34.

16. For *epitasis* as 'increase' cf. *dC* II 6 288a19; this is a more precise translation than Hett's (1936) 'crisis', or the Oxford Translation's 'climax' (G.R.T. Ross 1908); at 479a18 they give 'strain' and 'increase in strain', respectively, as renderings of the same word.

17. See the definition of suffocation in chapter 15: 'that kind of exhaustion due to lack of cooling is called suffocation' (475a28 f).

18. 'Violent disease' is Hett's translation for *biaion pathos* (479a21); one could take it more generally, since after all less violence (in the usual sense) may be needed to murder the old. Aristotle himself talks of the soul parting since, strictly speaking, it does not perish, but this does not mean that its activity continues. The parting of the soul (*apolysis* 479a22) is obviously a variant on Plato's definition of death in the *Phaedo* 67d9-10: the loosing and separation (*lysis kai chorismos*) of the soul from the body. As Hackforth (1955: 44 n 1) points out, this definition of death 'doubtless represents the normal contemporary view' and was accepted and adapted by both Stoics and Epicureans (cf. also *Phaedo* 64c4-8 for the escape (*apallagê*) of the soul, *Gorgias* 524b3 for its separation (*dialysis*) from the body); according to Aristotle (*JSVM* 10 471b30 ff (= DK 68 A106) esp. 472a14) Democritus had already defined death as the exit of the soul particles i.e. breath (cf. *Phaedo* 70a-b) from the body. The explanation of death in the *Timaeus* concentrates on the bodily events which cause the death of the mortal parts of the soul (*thymoeides, epithymêtikon*): on the one hand our bodies are unstable, since the elements they consist of tend towards their own places in the cosmos (81a), and on the other hand the triangles, responsible for working on food in bone-marrow, are sharp and tightly fitted together, but with the passage of time they become blunt, so causing the withering of the body in old age (81b-d). This provides Plato with his explanation of the life-cycle: each kind has its life-span, but then so does each individual, based on the fact that the triangles only have power to last a certain time (89b-c). Plato does not explain the connection between individual and specific life-span. It is, however, important to note the connection between death and the life-cycle, and between them and the function of nutrition; this last is, of course, not identified with a part of the soul. He also exploits the structure of matter in a way not open to Aristotle.

19. Cf. above p. 111. But surely that makes no sense, since on this reading the heat has grown less in the life of the individual, and yet it is the increase of this heat that is causing death? This paradox can be resolved as follows. First, we must remember that the exhaustion of a fire happens because the fire increases too much for the fuel available, and so dies out for lack of fuel. Now, a huge fire with enough fuel to keep it going would need a huge surge in heat for the fire to increase so much that it would consume its fuel so quickly that it would of a sudden go out. Now think of a match that suddenly flares and burns itself out. The flaring up of the match does not have to be that great for exhaustion to happen. So what Aristotle is suggesting is that in old animals there is very little heat left, and a small surge will exhaust it. And in fact, Aristotle offers us the comparison with a small flame that is easily put out by a slight movement (479a18-20).

20. 479a10-14, cf. chapter 24 479b2, above §4.2 p. 84 f.

21. Michael 1903:141.32-142.5 ad 478b26.

22. See above §4.1.

23. *GC* I 7, cf. above §4.2 on *LBV* 3.

24. 479a28, Hett's translation (1936).

25. See e.g. *GA* I 18 726a26, *PA* IV 2 677b8 (but cf. *GA* I 18 725a14 where phlegm seems to be a useful residue); cf. Peck 1953: Introduction, §§65-8.

26. 479a29-b3. At 479b2 I follow Siwek and Ross' reading *teleiotêta* 'completeness, fullness, maturity'. *Ph.* IV 13 222b25-7 make clear that time itself is not the agent; see above p. 84 f.

27. Note 'therefore' (*oun*) at 479a29.

28. Ogle's phrase (1897: 132 n 141) is 'first formation of the unfecundated germ' citing *GA* II 3 736a32 (followed by Ross 1955a: ad 479a29); but the unfecundated germ only has the capacity for nourishment, and not the actual activity (736b9). On the growth and development of the embryo of sanguineous animals, see *GA* II 4-6. Cf. *Ph.* V 1 225a12-20 on the termini of generation and perishing, quite generally and not just of living things.

29. The phrasing is difficult; I follow Ross' précis 1955a: 479a29, and his note ad loc.

30. Cf. the definition of life criticised by Aristotle at *Metaph.* VIII 6 1045b11-12 as the synthesis or combination of the soul with body (cf. *Top.* VI 14 151a21). Problems here are: (a) it is not clear what combination, or participation (1045b8-9, and cf. *Metaph.* I 6 987b9, XII 10 1075b19, VII 6 1031b18) means here; (b) a moving cause is needed to bring the soul about in the body (e.g. VIII 6 1045a30). Talk of participation, particularly in life, may of course be uncompromising, implying merely that something has life: *GA* II 1 732a12, 5 741a23, which Bonitz 1870 s.v. cites alongside our definition from *JSVM*; in *JSVM* 479a9 f heat shares (*koinônoun*) in life; for things 'having' life cf. *SS* 1 436a12. See the definitions of life using perception (*aisthêsis*), particularly in ethical contexts *EN* IX 9 1170a16, *EE* VII 12 1244b23, *Pol.* VII 1335b23-6 (abortion should happen before 'life and perception' are present).

31. Cf. *Rhet.* II 14 1390b6-9 on the prime of life as a mean, i.e. prey to neither the excesses or deficiencies of youth or old age. Cf. *dC* II 6 288a19, *dA* I 5 411a30, III 9 432b24, 12 434a24, *GA* IV 6 775a13. Prime and old age, caused by cold and heat, are attributed to the interior parts of the earth at *Meteor.* I 14 351a26.

32. These two kinds of definitions are the subjects of *APo.* II 8 and 9 respectively; primitive terms may well lie behind *APo.* II 9 (see Barnes ad loc. 1994: 221 f).

33. Matthews 1992 attempts to give a definition of life using the various functions listed in *dA* II 1-4 (esp. 413a22-5, 414a29-32); a single function, which Matthews understands as powers (cf. above §3.2) tending to preserve the species, would be enough for something to fulfil the definition. There are however, reasons for thinking that the functions listed in II 1-3 are only preliminaries, offered without explanation (cf. II 2 413a15).

34. 412a14 f.

35. 'First', from the definition of coming to be, is left out by Aristotle, quite reasonably: it is not the *first* participation in nutritive soul that remains, but the participation in nutritive soul.

36. 467b16-18; see above §3.5.

37. See above §3.5 for the arguments that there must be one central part responsible for nutrition and perception.

38. Cf. Barnes' view of induction (Barnes 1994: 264 on *APo.* II 19): 'We understand a notion A when all A rests in our minds; and that, presumably, will occur either when the successive impressions of past A's have etched a complete image in our minds, or else when we have enough concurrent images of A's to extract all and only the essential attributes of A.' The basic idea of induction is that we grasp the universal by attending to particular cases (*APo.* I 18 81b2). A first universal is recognised when a unitary species comes to a standstill: the individual is perceived, but as a member of a species. The next step lies in abstracting from several such species, until we arrive at what their genus is (*APo.* II 19 100a15-b5). For the difficulties in the account cf. Barnes 1994 ad loc. This procedure requires that the concepts concerned stand in relations of species and genera to one another: one is subordinate to another.

39. For the fuzziness of this border, due to plant-like animals, see Lloyd 1996: ch. 3 'Fuzzy Natures'.

40. See esp. chapter 1 467b27, 34, 468a4, 2 468a29, 3 468b16. In *PN*, there is a certain tension between the search for generality and the use of large blooded animals, exemplified in turn by humans, as a model (*SV* II 455b29-34). The idea of a centre is most obvious, if inexact, in such animals.

41. Chapter 4 469b5 f, 10 f, 17 f (cf. 3 469a11); 14 474b2-3.

42. Chapter 7 470b12-27; this passage interrupts the introduction of the doxographical part (chapters 8-13); thematically, in its discussion of cold animals with spongy lungs, it belongs with chapter 15 475a20-9, a passage which Ogle suspected of being misplaced (see above p. 109 n 206).

43. Chapter 15.

44. Chapter 16 475b15-26.

45. 475b27 f, 476b1 f.

46. The account here is meant to include plants (479b3-4: the end is called drying out in plants and death in animals). But there are difficulties, above all since plants have no organ which grows and decays, so causing the stages in their lives; instead, they rely on the environment and their food for cooling (*JSVM* 6). Part of the solution lies of course in the idea of the seasons (see above §4.2), but of course only part, since plants may also be perennial; on the longevity, particularly, of trees see *LBV* 6 (see above §4.3).

47. Cold is a lack of heat *GA* II 6 743a36; has its own nature *PA* II 2 649a18.

The closest he comes to defining heat in *PA* II is by saying there are two main senses: hot to the touch, and causing flame (649b4-5).

48. Cf. above §4.1.

49. The contrast is drawn at 479a29-30. Cf. death as a boundary (*peras*) at *EN* III 9 1115a26. Aristotle is not concerned with death as a *state*, opposed to the state of being alive. This contrasts with Plato's *Phaedo* (esp. 64c5-8), where the point is precisely that the state of being dead is preferable to the state of being alive.

50. *JSVM* 1 467b10.

51. On Aristotle's view of the heart as the first part that is formed and the last part that dies (*GA* II 5 741b19 ff), see Ebstein 1920: 305-6.

52. The idea of development plays a very important role in the relation between size and the length of gestation. On the eye's development, see *GA* II 6 743b32-744b11; on gestation, IV 4 772b6-8, 8 777a22, cf. above §4.2. Note also the importance of size in the account of feeding: feeding is connected with growth i.e. in size (see above §3.3).

Select Bibliography

Editions, translations and commentaries

References are generally made using the name of the editor, translator or commentator; texts used have not always been referred to using their editors' names.

Aristotle

Alexander of Aphrodisias. 1891. Commentaria in Aristotelis Metaphysica. In Hayduck, M. (ed.), *Commentaria in Aristotelem Graeca* vol. I. Berlin.
―――― 1899. Commentaria in Aristotelis Meteorologica. In: Hayduck, M. (ed.), *Commentaria in Aristotelem Graeca* vol. III pars ii. Berlin.
―――― 1901. In librum Aristotelis de Sensu commentarium. In Wendland, P. (ed.), *Commentaria in Aristotelem Graeca* vol. III pars i. Berlin.
―――― 1989. *On Aristotle Metaphysics 1*. Trans. W. Dooley. London.
Allan, D.J. 1965. *Aristotelis de Caelo Libri Quattuor*. Oxford: Oxford Classical Text.
Balme, D. 1992. *De Partibus Animalium I* and *De Generatione Animalium I* (with passages from II 1-3). Translated with Notes. Oxford: Clarendon Aristotle Series.
Barnes, J. (ed.) 1984. *The Complete Works of Aristotle. The Revised Oxford Translation*. 2 vols. Princeton.
―――― 1994. *Posterior Analytics*. Translated with a Commentary. second edition. Oxford: Clarendon Aristotle Series.
Bekker, I. 1831 (sqq.). *Aristotelis Opera*. Berlin.
Biehl, W. 1898. *Parva Naturalia*. Leipzig.
Bywater, I. 1894. *Aristotelis Ethica Nicomachea*. Oxford: Oxford Classical Text.
Charlton, W. 1970. *Aristotle's Physics I-II*. Translated with Introduction and Notes. Oxford: Clarendon Aristotle Series.
Drossaart Lulofs, H.J. 1965. *Aristotelis de Generatione Animalium*. Oxford: Oxford Classical Text.
Düring, I. 1944. *Aristotle's Chemical Treatise. Meteorologica Book IV*. Göteborg.
Flashar, H. 1962. *Aristoteles. Problemata Physica*. Aristoteles Werke in deutscher Übersetzung. Band 19. Berlin.
Frede, M. and Patzig, G. 1988. *Aristoteles 'Metaphysik Z'*. Text, Übersetzung und Kommentar. Munich.
Gallop, D. 1996. *Aristotle on Sleep and Dreams*. Peterborough, Ontario.
Hamlyn, D.H. 1993. *Aristotle De Anima. Books II and III* (with passages from Book I). Translated with Introduction and Notes. Oxford: Clarendon Aristotle Series.
Hett, W.S. 1936. *On the soul, Parva Naturalia, On breath*. Cambridge, Mass./London: Loeb Classical Library.
Hicks, R.D. 1907. *Aristotle: De Anima*. Cambridge.
Hussey, E. 1983. *Aristotle's Physics III-IV*. Translated with Notes. Oxford: Clarendon Aristotle Series.
Joachim, H.H. 1926. *Aristotle on coming to be and passing away*. Oxford.
Jaeger, W. 1957. *Aristotelis Metaphysica*. Oxford: Oxford Classical Text.

Select Bibliography

Kirwan, C. 1971. Aristotle's *Metaphysics*. Books Γ, Δ, E. Translated with Notes. Oxford: Clarendon Aristotle Series.

Lawson-Tancred, H. 1986. *Aristotle. De Anima (On the Soul)*. Translated with an Introduction and Notes. London.

Lee, H.D.P. 1952. *Aristotle, Meteorologica*. Cambridge, Mass./London: Loeb Classical Library.

Michael of Ephesus. 1903. Commentaria in Aristotelis Parva Naturalia In Wendland, P. (ed.), *Commentaria in Aristotelem Graeca* vol. XXII pars i. Berlin.

Minio-Paluello, L. 1949. *Aristotelis Categoriae et Liber de Interpretatione*. Oxford: Oxford Classical Text.

Nussbaum, M.C. 1978. *Aristotle's De Motu Animalium*. Princeton.

Ogle, W. 1897. *Aristotle on Youth and Age, Life and Death and Respiration*. London.

Peck, A.L. 1953. *Aristotle. Generation of Animals*. Cambridge, Mass./London: Loeb Classical Library.

—— 1961. *Aristotle. Parts of Animals, Movement of Animals, Progression of Animals*. Cambridge, Mass./London: Loeb Classical Library.

—— 1965-1991. *Aristotle, Historia Animalium*. 3 vols. Cambridge, Mass./London: Loeb Classical Library.

Philoponus, J. 1897a. In Aristotelis de anima libros tres. Hayduck, M. (ed.), *Commentaria in Aristotelem Graeca* vol. XV. Berlin.

—— 1897b. In Aristotelis libros de Generatione et Corruptione. Vitelli, H. (ed.), *Commentaria in Aristotelem Graeca* vol. XIV pars 2. Berlin.

Rodier, G. 1900. *Aristote: Traité de l'âme*. Traduit et annoté. 2 vols. Paris.

Ross, G.R.T. 1906. *Aristotle. De Sensu and De Memoria*. Cambridge.

—— and Beare, J.I. 1908. *Aristotle's Parva Naturalia*. Oxford: Oxford Translation, vol. III.

Ross, W.D. 1924. *Aristotle's Metaphysics*. Oxford.

—— 1928. *Metaphysica*. Oxford: Oxford Translation, vol. VIII.

—— 1936. *Aristotle's Physics*. Oxford.

—— 1955a. *Aristotle's Parva Naturalia*. Oxford.

—— 1955b. *Aristotelis Fragmenta Selecta*. Oxford: Oxford Classical Text.

—— 1957. *Aristotelis Politica*. Oxford: Oxford Classical Text.

—— 1958. *Aristotelis Topica et Sophistici Elenchi*. Oxford: Oxford Classical Text.

—— 1959. *Aristotelis Ars Rhetorica*. Oxford: Oxford Classical Text.

—— 1961. *Aristotle's de Anima*. Oxford.

—— 1964. *Aristotelis Analytica Priora et Posteriora*. Oxford: Oxford Classical Text.

Simplicius. 1882. *In Aristotelis Physicorum Libros Quattuor Priores Commentaria*. Diels, H. (ed.). Berlin.

Siwek, P. 1963. *Aristotelis Parva Naturalia graece et latine*. Rome.

Sorabji, R. 1972. *Aristotle on Memory*. London.

Thompson, D. 1910. *History of Animals*. Oxford: Oxford Translation, vol. IV.

Trendelenburg, F. 1877. *Aristotelis de anima libri tres*. Berlin.

Tricot, J. 1951. *Parva Naturalia*. Paris.

Van der Eijk, P.J. 1994. *de Insomniis, de Divinatione per Somnium*. Aristoteles Werke in deutscher Übersetzung. Band 14, Teil III. Berlin.

Wardy, R. 1990. *The Chain of Change. A Study of Aristotle's Physics VII*. Cambridge.

Williams, C.J.F. 1982. *Aristotle's De Generatione et Corruptione*. Translated with Notes. Oxford: Clarendon Aristotle Series.

Other ancient authors

Bruns, I. 1887. Alexandri Aphrodisiensis praeter Commentaria scripta minora. de Anima cum Mantissa. *Supplementum Aristotelicum* Vol. II pars I. Berlin.

Burnet, J. 1900-1907. *Platonis opera*. 5 vols. Oxford.

Diels, H. and Kranz, W. 1951. *Die Fragmente der Vorsokratiker*. 3 vols. Sixth edition. Berlin.

Fortenbaugh, W.W., Huby, P.M., Sharples, R.W. and Gutas, D.M. (ed. and trans.), *Theophrastus of Eresus, Sources for his Life, Writings, Thought and Influence*. Vol. 5: R.W. Sharples,

Select Bibliography

Sources on Biology. Text 1992; Commentary 1995. Vol. 3.1: *Sources on Physics.* Commentary. 1998. Leiden.
Gottschalk, H.B. 1965. Strato of Lampsacus: some texts. *Proceedings of the Leeds Philosophical and Literary Society, Literary and Historical Section* 11.2, 95-182.
Hackforth, R. 1955. *Plato's Phaedo Translated with an Introduction and Commentary.* Cambridge.
Kirk, G.S. 1962. *Heraclitus. The Cosmic Fragments.* Cambridge.
Kühn, C.G. 1821-33. *Claudii Galeni Opera Omnia.* Leipzig, repr. Hildesheim 1965.
Littré, E. 1839-61. *Oeuvres complètes d'Hippocrate.* 10 vols. Paris.
Marcovich, M. 1967. *Heraclitus.* Merida.
Martin, J. 1992. *T. Lucretius Carus. De rerum natura.* Stuttgart.
Muehll, P. von der. 1922. *Epicuri epistulae tres et ratae sententiae.* Leipzig.
Ruland, H.J. 1976. Die arabischen Fassungen von zwei Schriften des Alexander von Aphrodisias über die Vorsehung und über das liberum arbitrium. Diss. Saarbrücken.
Sharples, R. 1983. *Alexander of Aphrodisias On Fate.* London.
Stratton, G.M. 1917. *Greek Physiological Psychology.* London (containing Theophrastus' *de Sensibus*).
Todd, R.B. 1976. Alexander of Aphrodisias on Stoic Physics. *Philosophia antiqua* XXVIII. Leiden (= *de mixtione*).
West, M.L. 1971. *Iambi et Elegi ante Alexandrum cantati.* 2 vols. Oxford.
Wimmer, F. 1854-62. *Theophrasti Eresii opera.* Leipzig.

Modern authors

Where applicable, articles are cited from the reprinted version. References to a modern author may be to a modern edition, commentary or translation of an ancient text.

Ackrill, J. 1972. Aristotle's definition of psuchê. *Proceedings of the Aristotelian Society* 73, 1972-3, 119-33. Also in Barnes, J. et al. (edd.) 1975-79, vol. 4, 65-75.
Althoff, J. 1991. Warm, Kalt, Flüssig Fest bei Aristoteles. *Hermes Einzelschriften* 57. Stuttgart.
—— 1992. Das Konzept der generativen Wärme bei Aristoteles. *Hermes* 120 (2), 181-93.
—— 1997. Aristoteles Vorstellung von der Ernährung der Lebewesen. In Kullmann and Föllinger 1997, 351-64.
—— 1999. Aristoteles als Medizindoxograph. In Van der Eijk 1999, 57-94.
Anscombe, G.E.M. 1979. The principle of individuation. In Barnes, J. et al. (edd.) 1975-79, vol. 3.
Balme, D. 1962. Aristotle's use of differentiae in zoology. In Barnes, J. et al. (edd.) 1975-79, vol. 1, 183-93.
—— 1987a. Aristotle's use of differentiae in zoology (revised version). In Gotthelf, A. and Lennox, J.G. (edd.) 1987, 69-89.
—— 1987b. Teleology and necessity. In Gotthelf, A. and Lennox, J.G. (edd.) 1987, 275-85.
—— 1987c. The place of biology in Aristotle's philosophy. In Gotthelf, A. and Lennox, J.G. (edd.) 1987, 9-20.
Barker, A. 1981. Aristotle on perception and ratios. *Phronesis* 26, 148-66.
Barnes, J., Schofield, M. and Sorabji, R. (edd.) 1975-79. *Articles on Aristotle.* 4 vols. London.
Bogen, J. 1995. Fire in the belly: Aristotelian elements, organisms, and chemical compounds. *Pacific Philosophical Quarterly* 76, 370-404.
Boll, F. 1950. Die Lebensalter: ein Beitrag zur antiken Ethologie und zur Geschichte der Zahlen. In Boll, F., *Kleine Schriften zur Sternkunde des Altertums.* Leipzig.
Bonitz, H. 1870. *Index Aristotelicus.* Berlin.
Boylan, M. 1982. The digestive and 'circulatory' systems in Aristotle's biology. *Journal of the History of Biology* 15, 89-118.
Brady, J. 1979. *Biological Clocks.* London.
Broadie, S. 1990. Nature and craft in Aristotelian teleology. In Devereux and Pelegrin 1990, 389-403. See also under Waterlow.

Select Bibliography

Browning, R. 1962. An unpublished funeral oration on Anna Comnena. *Proceedings of the Cambridge Philological Society* n.s. 8, 1-12. Reprinted in Sorabji, R. (ed.) 1990, 393-406.

Buchheim, Th. 1999. Vergängliches Werden und sich bildende Form. Überlegungen zum frühgriechischen Naturbegriff. *Archiv für Begriffsgeschichte* 41, 7-34.

—— 2001. The functions of the concept of *physis* in Aristotle's *Metaphysics*. *Oxford Studies in Ancient Philosophy*.

Burnyeat, M.F. (ed.). 1979. *Notes on Book Z of Aristotle's Metaphysics*. Oxford.

—— 1980. Aristotle on learning to be good. In Rorty 1980, 69-92.

—— 1992. Is an Aristotelian philosophy of mind still credible? (a draft). In Nussbaum, M.C. and Rorty, A.O. (edd.) 1992, 15-26.

Byl, S. 1980. Recherches sur les grands traités biologiques d'Aristote: sources, écrits et préjugés. *Academie Royale de Belgique, Mémoires de la Classe des Lettres* LXIV Fasc. 3.

Cassirer, H. 1932. *Aristoteles Schrift 'Von der Seele'*. Tübingen.

Charles, D. 1988. Aristotle on hypothetical necessity and irreducibility. *Pacific Philosophical Quarterly* 69, 1-53.

—— 1991. Teleological causation in the *Physics*. In Judson 1991.

Code, A. 1995. Potentiality in Aristotle's science and *Metaphysics*. *Pacific Philosophical Quarterly* 76, 405-18.

—— and Moravcsik, J. 1992. Explaining various forms of living. In Nussbaum and Rorty 1992.

Comfort, A. 1979. *The biology of Senescence*. 3rd ed. London.

Cooper, J. 1982. Aristotle on natural teleology. In Schofield, M. and Nussbaum, M.C. (edd.) 1982, 197-222.

—— 1985. Hypothetical necessity. In Gotthelf, A. (ed.) 1985, 151-67.

—— 1987. Hypothetical necessity and natural teleology. In Gotthelf, A. and Lennox, J.G. (edd.) 1987, 243-74.

Davies, I. 1983. *Ageing*. The Institute of Biology's Studies in Biology no. 151. London

Debru, A. 1996. *Le corps respirant. La pensée physiologique chez Galien*. Leiden.

Devereux, D. and Pellegrin. P. (ed.). 1990. *Biologie, Logique et Métaphysique chez Aristote*. Paris.

Dilcher, R. 1995. *Studies in Heraclitus*. Spudasmata Band 56. Hildesheim.

Ebstein, E. et al. 1920. Woher stammt und wo kommt zuerst vor: cor ultimum moriens? *Mitteilungen zur Geschichte der Medizin und der Naturwissenschaft* 19: 102, 219, 305-6.

Emson, H.E. 1995. Health, aging and death. *Cambridge Quarterly of Healthcare Ethics* 4, 163-6.

Everson, S. 1997. *Aristotle on Perception*. Oxford.

Feldman, F. 1992. *Confrontations with the Reaper. A Philosophical Study of the Nature and Value of Death*. New York.

Fine, K. 1995. The problem of mixture. *Pacific Philosophical Quarterly* 76, 266-369.

Fischer, J.M. (ed. with intro.). 1993. *The Metaphysics of Death*. Stanford.

Föllinger, S. 1996. Gleichheit und Differenz. Das Geschlechterverhältnis in der Sicht griechischer Philosophen des 4. bis 1. Jahrhunderts vor Christi. *Hermes Einzelschriften* 74. Stuttgart.

Frede, M. 1980. The original notion of a cause. In M. Schofield, M. Burnyeat and J. Barnes (edd.). *Doubt and Dogmatism: Studies in Hellenistic Epistemology*. Oxford. Reprinted in M. Frede, 1987, *Essays in Ancient Philosophy*. Oxford.

—— 1990. The definition of sensible substances in Metaphysics Z. In Devereux and Pellegrin 1990, 287-320.

Freeland, C. 1990. Aristotle on bodies, matter and potentiality. In Devereux and Pellegrin 1990, 392-407.

Freudenthal, G. 1995. *Aristotle's Theory of Material Substance*. Oxford.

Freudenthal, J. 1869. Zur Kritik und Exegese von Aristoteles *Peri tôn koinôn sômatos kai psychês ergôn* (Parva Naturalia). *Rheinisches Museum* 24, 81-93, 392-419.

Furley, D.J. 1984. Theories of respiration before Galen. In Furley, D.J. and Wilkie, J.S., *Galen on Respiration and Arteries*. Princeton, 3-39.

Garland, R. 1990. *The Greek Way of Life: From Conception to Old Age*. London.

Select Bibliography

Gill, M.L. 1989a. Aristotle on matters of life and death. *Boston Area Colloquium in Ancient Philosophy* 4, 187-205.

—— 1989b. *Aristotle on Substance: The Paradox of Unity*. Princeton.

Gotthelf, A. (ed.) 1985. *Aristotle on Nature and Living Things*. Pittsburgh.

—— 1989. The place of the good in Aristotle's natural theology. *Boston Area Colloquium in Ancient Philosophy* 5, 113-39.

—— and Lennox, J.G. (edd.) 1987. *Philosophical Issues in Aristotle's Biology*. Cambridge.

Guariglia, Osvaldo. N. 1985. Die Definition und die Kausalerklärung bei Aristoteles. In Menne and Offenberger 1985.

Harris, C.R.S. 1973. *The Heart and the Cardio-vascular System in Antiquity*. Oxford.

Healy, J.F. 1978. *Mining and Metallurgy in the Ancient World*. London.

Heidegger, M. 1967. Vom Wesen und Begriff der *physis*. Aristoteles, Physik B, 1. In Heidegger, M. 1967. *Wegmarken*. Frankfurt am Main, 237-99.

Hintikka, J. 1973. *Time and Necessity*. Oxford.

Hübner, J. 1999. Die Aristotelische Konzeption der Seele als Aktivität. *Archiv für die Geschichte der Philosophie* 81, 1-32.

—— 2000. *Aristoteles über Getrenntheit und Ursächlichkeit. Der Begriff des eidos chôriston*. Hamburg.

Hussey, E. 1991. Aristotle's mathematical physics: a reconstruction. In Judson, L. (ed.) 1991.

Hutchinson, D.S. 1987. Restoring the order of Aristotle's *de Anima. Classical Quarterly* 37, 373-81.

Jaeger, W. 1913. Das Pneuma im Lykeion. *Hermes* 48, 29-74.

Joachim, H. 1904. Aristotle's conception of chemical combination. *Journal of Philology* 29, 72-86.

Johansen, T.K. 1998. *Aristotle on the Sense Organs*. Cambridge.

Judson, L. (ed.) 1991. *Aristotle's Physics: A Collection of Essays*. Oxford.

Kahn, C. 1966. Sensation and consciousness in Aristotle's psychology. In Barnes, J. et al. (edd.) 1975-79, vol. 4, 1-31.

—— 1988. From the philosophy of being to the philosophy of human being. In D. Henrich and R.P. Horstmann (edd.) 1988. *Metaphysik nach Kant*. Stuttgart, p. 528-40.

Kant, I. 1902 f. *Kritik der Urteilskraft*. Kants Werke. Akademie-Textausgabe vol. 5. Berlin.

Keyt, D. 1989. The meaning of *bios* in Aristotle. *Ancient Philosophy* 9, 15-21.

King, R.A.H. 1998. Making things better. The art of changing things. (Aristotle *Metaphysics* IX 2) *Phronesis* 43 (1), 63-83.

Kosman, L.A. 1969. Aristotle's definition of motion. *Phronesis* 14, 40-62.

—— 1984. Substance, being and *energeia. Oxford Studies in Ancient Philosophy* 2, 121-49.

—— 1994. The activity of being in Aristotle's *Metaphysics*. In Scaltsas, Charles and Gill 1994, 195-213.

Kullmann, W. 1979. *Die Teleologie in der aristotelischen Biologie*. Sitzungsberichte der Heidelberger Akademie der Wissenschaften. Philosophisch-historische Klasse. Heidelberg.

—— 1985. Notwendigkeit in der Natur bei Aristoteles. In Wiesner 1985: 207-38.

—— and Föllinger, S. (edd.) 1997. *Aristotelische Biologie. Intentionen, Methoden, Ergebnisse*. Stuttgart.

Le Blond, J.M. 1938. *Eulogos et l'argument de convenance chez Aristote*. Paris.

—— 1979. Aristotle on definition. In Barnes, et al. (edd.) 1975-79, vol. 3, 63-79.

Lennox, J. 1997. Nature does nothing in vain. In *Beiträge zur antiken Philosophie. Festschrift für Wolfgang Kullmann*, ed. H.-Chr. Günther and A. Rengakos. Stuttgart, 199-214.

Lewis, F.A. 1994. *Aristotle on the relation between a thing and its matter*. In Scaltsas, Charles and Gill 1994, 247-79.

Lloyd, G.E.R. 1966. *Polarity and Analogy*. Cambridge.

—— 1973. *Greek Science after Aristotle*. Cambridge.

—— 1975. Alcmaeon on the early history of dissection. *Sudhoffs Archiv* 59, 113-47. Reprinted in Lloyd 1991, 164-93.

—— 1978. The empirical basis of the physiology of the *Parva Naturalia*. In Lloyd, G.E.R. and Owen, G.E.L. (edd.) 1978, 215-39. Reprinted in Lloyd 1991, 224-47.

199

Select Bibliography

—— 1983. *Science, Folklore and Ideology*. Cambridge.

—— 1991. *Methods and Problems in Greek Science*. Cambridge. (All page references to articles are to this volume.)

—— 1992. Aspects of the relationship between Aristotle's psychology and his zoology. In Nussbaum, M.C. and Rorty, A.O. (edd.) 1992, 147-67.

—— 1996. *Aristotelian Explanations*. Cambridge.

—— and Owen, G.E.L. (edd.) 1978. *Aristotle on Mind and the Senses*. Proceedings of the Seventh Symposium Aristotelicum. Cambridge.

Louis, P. 1952. Le traité d'Aristote sur la nutrition. *Revue de Philologie* 26, 29-35.

Mansfeld, J. 1967. Heraclitus on the psychology and physiology of sleep and on rivers. *Mnemosyne* IV 20, 1-29.

Mansion, S. 1978. Soul and life in the *de Anima*. In Lloyd and Owen (edd.) 1978, 1-20.

Matthews, G.B. 1992. *De Anima* 2.2-4 and the Meaning of *Life*. In Nussbaum and Rorty (edd.) 1992.

Menne, A. and Offenberger, N. (ed.) 1982, 1985, 1988, 1990. *Zur modernen Deutung der Aristotelischen Logik*. 4 vols. Hildesheim.

Mercken, H.P.F. 1973. *The Greek Commentators on Aristotle's Ethics*. Introduction to Corpus Latinum Commentatorium in Aristotelem Graecorum. Leiden. vol. 6:*3-*29. Reprinted in Sorabji (ed.) 1990, 407-44.

Mertens-Horn, M. 1997. Il trono Ludovisi e il trono di Boston. Rappresentazione di scene sacre. *Quaderni di Palazzo Grassi*, 94-106.

Modrak, D. 1987. *Aristotle. The Power of Perception*. Chicago.

Moraux, P. 1951. *Les listes anciennes des ouvrages d'Aristote*. Louvain.

Moraux, P. and Wiesner, J. (edd.) 1983. *Zweifelhaftes im Corpus Aristotelicum*. Berlin.

Nagel, Th. 1979. Death. In his *Mortal Questions*. Cambridge, 1-10.

Neuhäuser, J. 1878. *Aristoteles' Lehre von dem sinnlichen Erkenntnisvermögen und seinen Organen*. Leipzig.

Nussbaum, M.C. and Putnam, H. 1992. Changing Aristotle's mind. In Nussbaum M.C. and Rorty, A.O. (edd.) 1992, 27-56.

—— and Rorty, A.O. (edd.) 1992. *Essays on Aristotle's De Anima*. Oxford.

Nuyens, F. 1948. *L'evolution de la psychologie d'Aristote*. Louvain.

Owen, G.E.L. 1961. 'Tithenai ta phainomena'. In Mansion, S. (ed.) 1961. *Aristote et les problèmes de la méthode*. Louvain, 83-103. Reprinted in Barnes, J. et al. (edd.) 1975-79, vol. 1, 113-26.

Pellegrin, P. 1982. *La classification des animaux chez Aristote*. Paris. (English translation: *Aristotle's classification of animals*, translated by A. Preus. 1986. Berkeley.)

Platt, A. 1921. Aristotle on the heart. In Singer, Ch. (ed.), *Studies in the History of Science*, vol. 2, Oxford, 521-32.

Preus, A. 1990. Man and cosmos in Aristotle. In Devereux and Pellegrin 1990.

Rorty, A.O. (ed.) 1980. *Essays on Aristotle's Ethics*. Berkeley.

Rüsche, F. 1930. *Blut, Leben und Seele. Ihr Verhältnis nach Auffassung der griechischen und hellenistischen Antike, der Bibel und der alten Alexandrin- ischen Theologen*. Studien zur Geschichte und Kultur des Altertums. Supplement 5. Paderborn.

Scaltsas, T., Charles, D. and Gill, M.L. (edd.) 1994. *Unity, Identity and Explanation in Aristotle's Metaphysics*. Oxford.

Sisko, J.E. 1996. Material alteration and cognitive activity in Aristotle's *de Anima*. *Phronesis* 41, 138-57.

Solmsen, F. 1957. The vital heat, the inborn pneuma and the aether. *Journal of Hellenic Studies* 72: 119-23.

—— 1968. Cleanthes or Posidonius? The basis of Stoic Physics. In F. Solmsen, *Kleine Schriften*. Hildesheim.

Sorabji, R. 1964. Function. *Philosophical Quarterly* 14.

—— 1974. Body and soul in Aristotle. *Philosophy* 49: 63-89. Reprinted in Barnes, J. et al. (edd.) 1975-79, vol. 4, 42-64.

—— 1980. *Necessity, Cause, and Blame: Perspectives on Aristotle's Theory*. London.

Select Bibliography

———— 1990 (ed.). *Aristotle Transformed: The Ancient Commentators and their Influence.* London.

———— 1992. Intentionality and physiological processes: Aristotle's theory of sense perception. In Rorty, A.O. and Nussbaum, M.C. (edd.) 1992, 195-226.

Sourvinou-Inwood, C. 1995. *'Reading' Greek Death. To the End of the Classical Period.* Oxford.

Sprague, R. 1977. Aristotle and the metaphysics of sleep. *Review of Metaphysics* 31, 230-41.

Strohm, H. 1983. Beobachtungen zum vierten Buch der Aristotelischen Meteorologie. In Moraux, P. and Wiesner, J. (edd.) 1983, 94-115.

Thielscher, P. 1948. Die relative Chronologie der erhaltenen Schriften des Aristoteles nach den bestimmten Selbstzitaten. *Philologus* 97, 229-65.

Torraca, L. 1958. Sull' autenticità del de Motu Animalium di Aristotele. *Maia* 10: 220-33.

Tracy, T. 1969. *Physiological Theory and The Doctrine of the Mean in Plato and Aristotle.* Chicago.

———— 1983. Heart and soul in Aristotle. In Anton, J.P. and Preus, A. (edd.) 1983. *Essays in Ancient Greek Philosophy* ii, Albany, N.Y., 321-39.

Van der Eijk, P. 1997. The matter of mind: Aristotle on the biology of 'psychic' processes and the bodily aspect of thinking. In Kullmann and Föllinger 1997, 231-58.

———— 1999. *Ancient Histories of Medicine. Essays in Medical Doxography and Historiography in Classical Antiquity.* Studies in Ancient Medicine, vol. 20, ed. J. Scarborough. Leiden.

Verdenius, W.J. and Waszink, J.H. 1946. *Aristotle. On coming to be and passing away. Some Comments.* Leiden.

Waterlow, S. 1982. *Nature, Change and Agency in Aristotle's* Physics. Oxford.

Webb, P. 1982. Bodily structure and psychic faculties in Aristotle's theory of perception. *Hermes* 110, 25-50.

Wieland, W. 1962. *Die Aristotelische Physik.* Göttingen.

Wiesner, J. 1978. The unity of the treatise de Somno and the physiological explanation of sleep in Aristotle. In Lloyd, G.E.R. and Owen, G.E.L. (edd.) 1978, 241-80.

———— 1985. Aristoteles. *Werk und Wirkung.* Berlin.

Wiggins, D. 1980. *Sameness and Substance.* Oxford.

Williams, B. 1993. The Makropoulos Case: Reflections on the Tedium of Immortality. In Fischer 1993, 71-92.

Woodfield, A. 1976. *Teleology.* Cambridge.

201

Index locorum

General index

For Greek and Roman authors see the Index locorum. References of the form 66n180 may mean that the topic is discussed in the main text on p. 66 as well as in the note 180 referred to on that page. ')(' indicates a contrast of some kind. Greek words are translated; references are listed under the English entry.

210

www.ingramcontent.com/pod-product-compliance
Lightning Source LLC
Chambersburg PA
CBHW050354030726
47503CB00006B/1852